How Humanity Came Into Being

Martin Lockley
with Ryo Morimoto

How Humanity
Came Into Being

The Evolution of Consciousness

Floris Books

To Alyssa

First published in 2010 by Floris Books

Text © 2010 Martin Lockley and Ryo Morimoto
Illustrations © 2010 Martin Lockley

Martin Lockley and Ryo Morimoto have asserted their right
under the Copyright, Designs and Patent Act 1988 to be
identified as the Authors of this Work.

British Library CIP Data available
ISBN 978-086315-732-5
Printed in Glasgow by Bell & Bain Ltd.

Contents

Preface and Acknowledgments

This book has something of a dual origin. The first impetus to the senior author's interest in consciousness studies began in the 1980s and 1990s as a result of various experiences, meetings and discussions with like-minded thinkers. The second impetus to the book's origin has been a course taught at the University of Colorado, Denver, between 1998 and the present time (2010). This course, taught by the senior author, and initially entitled *The Evolution of Consciousness*, was later offered under the similar title *The Evolution of Consciousness and Culture*. Students who took the class, including the junior author, could at various times take the class for Geology, Anthropology, Philosophy or Religious studies credit. We therefore thank those students and faculties from these various programs who had faith in the value of such an interdisciplinary course and thereby helped to generate interest. They also gave us valuable feedback over many years.

These colleagues include Candice Shelby, Ph.D., Chair of the Philosophy Department, University of Colorado at Denver, who was also kind enough to read the draft manuscript, and Sharon Coggan, Ph.D., the indefatigable, and very popular Director of the University of Colorado at Denver Religious Studies Program, who has done her best to instil a rational scholarly approach to Religious Studies in those of us inclined to disorganized and tangential thinking on the subject. I am sure she will not mind us revealing that her religious inclination is towards evolutionary Rock and Roll ('Praise the Lord!' — her quote)!

The manuscript has benefited enormously from having been read by several friends and colleagues including Regula Noetzli, Pine Plains, New York; Joe Tempel, Morrison, Colorado; Bill Peterson Ph.D., University of Colorado at Boulder, who is an actual, proverbial 'rocket scientist'; and Steve McIntosh, Boulder, Colorado, a rising star in the field of Consciousness Studies. Last but by no means least we thank David Lorimer, Programme Director of the Scientific and Medical

Network, UK, who with his writing, editing, symposium organizing, and above all his impressive and profound grasp of a wide range of subjects, has been an inspiration for some years.

We also wish to thank Wolfgang Schad, Ph.D., Institute of Evolutionary Biology and Morphology, University of Witten-Herdecke, Witten, Germany, for his inspiration and for sharing with us his extraordinary holistic understanding of biology. Other colleagues who have provided advice, encouragement and valuable discussion of holistic 'Schadian' biology, include Craig and Henrike Holdredge, The Nature Institute, Ghent, New York; Ken McNamara, Ph.D., Department of Earth Sciences, Cambridge University, UK; Mark Riegner, Ph.D., Prescott College, Arizona; and Bernd Rosslenbroich, Ph.D., also of the Institute of Evolutionary Biology and Morphology, University of Witten-Herdecke, Witten, Germany. We also thank Luis Ferreyra for his help with much of the artwork, and Helena Waldron for her work on the layout.

We thank our editor, Christopher Moore, for his skilful and valuable advice, and we also thank Christian Maclean and the crew at Floris Books, Edinburgh, for their assistance and support. We both thank our families — the Lockleys and the Morimotos — for their ongoing support and assistance. Finally we offer our very special thanks and gratitude to the Heberton family for their strength, kindness, generosity and support during difficult trials in recent years.

Introduction

The tranquillity of the morning meditation is interrupted by an extraordinarily powerful sensation of energy rising within the body. It begins like an internal tingling, but soon feels like a powerful electric current arising from nowhere and everywhere all at once. At first the stimulation intrigues, as a new physical sensation, but in no time the feeling becomes both disconcerting and awe-inspiring. The emotions and mind clamour to make sense of the uncontrolled tidal wave of energy. Is the body-mind unexpectedly losing control? But it is impossible to 'do' anything to respond to the electric storm brewing inside. The lightning strikes and the whole world is illuminated within and without. There is no time to wonder whether to ride the wave. Reality and meaning are shifting fast and the 'wonder' is of an entirely different kind — deeply experiential. The wave crest is already reached, and it is luminous, psychedelic, transcendent, divine.

Reflecting on this consciousness experience a year later, our protagonist who during the episode was mute, literally struck dumb, now has a few, still meagre words to describe the epiphany. The energy that surged up the spine, to emerge like the bursting of a luminous lotus bud, has been called the rising of the kundalini (meaning 'coiled serpent').[1] It is a comparatively well known experiential phenomenon in some yogic traditions, though still little understood in the west. What resonates most deeply with our protagonist, who now has a much deeper interest in authentic yoga, is that the kundalini episode was life-changing. Though the intensity of the immediate experience faded, there remains a permanent sense of expanded consciousness. A primal well of evolutionary energy erupted as pure consciousness, searing its way up from unconscious magmatic depths.

Compare this description with the following types of experience: a
child chortles contentedly in his cradle, snug in the soothing, sensory
caress of dappled sunshine flickering through the shimmering leaves of
the trees overhead. The trembling foliage, radiant sunshine and happy
gurgles merge with the soporific buzzing of bees to whisper 'unity,
unity, unity ...' The infant, ripe orchard and sun-bathed planet languish
harmoniously in timeless garden-of-Eden-consciousness. The biosphere
buzzes its eternal, unconscious lullaby of life.

A young athlete sprints down the runway towards the long jump pit
and launches from the springy, take-off board with perfect precision.
By a miracle his first kick in mid-air imparts a pleasant sensation of
weightlessness, and he senses he will jump further than ever before.
Another well timed kick and his flight is extended. Incredible as it
seems he is consciously aware of an almost superhuman ability to stay
aloft, soaring on and on. Like a heroic Olympian his pedaling legs carry
him in exuberant, time-dilated flight towards a world record leap. Then
he awakens, still acutely aware that this particular dream, one he's had
several times before, seems completely real. Surely it could still be real
and yet happen in real life?

A car horn honks and a middle-aged woman realizes that for the last
few minutes she has not been consciously aware of driving home on her
daily commute. She must have 'tuned out' at least five miles ago. But,
even though she has no recollection of any distracting thoughts, it all
seems natural enough. She is not afraid of not having had her mind on
the business of driving. Sometimes one is aware of time and the world
around; sometimes one is not. While paying more conscious attention
to the road ahead, she wonders: 'What is consciousness?' Perhaps she
should explore the subject more deeply.

These four episodes serve as our introduction to the spectrum of
consciousness of humanity. They are but small manifestations of the
collective, infinitely-varied and endlessly dynamic human experience.
This then is consciousness — our ever-present participation in the
experience of life.

It seems to us that consciousness is inextricably interwoven with
organic existence and may very well be a manifestation of the primary

datum that structures the very fabric of the cosmos. Or, put another way, this is as good an explanation as any we might conceive. Many awestruck cosmologists have waxed lyrical on this very subject. As sentient beings we have as much chance of stepping out of the interwoven cosmic field as we have of single-handedly creating a separate universe. As philosophers like Spinoza have pointed out, the structure of the psyche resonates with the structure of the world.[2] Consciously and unconsciously we are embedded in the psychic, organic web of the biosphere. The web is already more than three billion years old, and infinitely intricate. We are part of it both physically and psychologically.

By most reasoned accounts the only way we have reached this conclusion is by becoming self-conscious and gaining the ability to reflect on who we are, first as individuals and then as a species. Our individual consciousness and that of the cultures in which we are embedded is so familiar that we find it difficult to see the world any other way. We regard our conscious experience as normal. But paradoxically, it is this very self-consciousness that allows us to appreciate the 'other'. Were this not the case, were we to live in an unawakened state of consciousness, we would still be part of the living biosphere, like all sentient creatures, including our ancestors, but ...

But, it would all be different. We could not perceive the world as we think we know it. There would be little or no sense of any relationship to space and time, so no conscious sense of relationships to other animals, other people, other things, no sense of history or of being embedded in evolutionary time. Hard to imagine? Perhaps, but, in essence, such experience is no different from our innocent, non-self-conscious state of perception as infants. Life, it appears, was once like this. In fact life has been like this for most of history and mostly remains this way. How soon we have forgotten!

But wait! Surely this talk of forgetting makes no sense whatsoever. Isn't it quite impossible to remember what it was once like not to remember? Or can we penetrate these mists of time to 'see' into unknown, psychic realms, past lives, the mind of Australopithecus or the dinosaurs? Some claim such glimpses. But although many are sceptical of the 'reality' of such purported insight, it is vastly intriguing that, as a species, we follow the same paths back into the unknown using both intellectual exploration of empirical evidence and abstract ideas about the past. We have awoken collectively to memory and historical time and can begin to consciously explore how the unconscious memory of the biosphere operated for all those billions of years. As we build our creation stories and scientific theories of origins and history, we are doing nothing short of making an archaeological excavation of consciousness, shedding the light of waking consciousness on the subconscious and unconscious realms that preceded our present state.[3] Isn't the paleontological excavation of our mammalian roots and deep-time cellular origins a manifestation of archetypal 'group memory'? Is it so different, in principle, from a past life memory? Whatever it is, it is undeniably a quest deeply woven into the fabric of consciousness and an indication of our profound desire for self-knowledge and a collective understanding of our species origins and place in the universe.

If this journey in memory and cosmic space-time is really happening, as growing evidence suggests, we can logically argue that we are still barely awake. Only the geniuses and illuminati among us can penetrate the mists of time to shed light on the obscurity of the unconscious. Legend tells us that the Buddha did precisely this, consciously penetrating back to the dawn of evolution. But these gifted persons are not mere individuals, they are part of human collective consciousness. We are all in this together. As one awakens followed by others, so the biosphere begins to awaken. The slumbering giant no longer slumbers. The lurching sleepwalker can begin to take deliberate strides. And the man-in-the-street can talk, like Sigmund Freud or Carl Jung, of the 'unconscious'. If it is really true that the biosphere and evolution itself is becoming conscious, and emerging from the unconscious abyss, these are exciting times. We are experiencing a consciousness growth spurt. Nothing could be more natural and organic. Evolution has been characterized by many periods of acceleration. The mode and tempo change — in evolutionary jargon, cycles of expansion (growth spurts) punctuate periods of equilibrium.

In a few short millennia that culminate three million years of hominid prehistory, and three billion years of evolution of sentient life, our species has articulated creation myths and invented philosophy, religion and science.[4] In much the same organic order in which individuals develop their physical, emotional and mental/spiritual faculties, the last few centuries have witnessed the sequential emergence of the physical, biological and psychological sciences. Our interest in consciousness, whether traced back to shamanism, the Hindu ancients, Freud, Jung or contemporary consciousness studies institutes, is in itself symptomatic of the biosphere's awakening.

'Hello, consciousness' is, therefore, an appropriate salutation for each new sentient birth and for each new cultural awakening or enlightenment. Since our peculiar brand of self-consciousness is so tied up with the concept of time, we adopt a historical framework in our discourse on the evolution of consciousness. Our science tells us that the physical, inorganic universe came into being some 13–15 billion years ago, and that the organic biosphere sprouted on Earth's surface some 3.5 billion years ago. Supposedly, it was not until the Axial Age, less than three thousand years ago, that humans self-consciously recorded their interest in mind, philosophy, religion and spirituality. Collectively we might label these human phenomena as 'psychology' (from the Greek *psyche,* meaning 'butterfly'). So in the best academic traditions, we see evolution as a progression from physical to biological, mental and spiritual. As just noted, it strikes us as more than mere coincidence that the emergence of the scholarly disciplines of physics/chemistry, biology and psychology in post-Renaissance Europe should march so sequentially from the seventeenth through the eighteenth, nineteenth and twentieth centuries to what has, in some circles, been called the 'post-religious spirituality' of the twenty-first century.[5] Since the normal development of the individual involves mastering the physical body, before coming to terms with his/her biological (emotional and social) and mental faculties, we are confident that the evolution of the universe, human culture and the individual follow similar or resonant, recapitulating trajectories that are either inherent in the dynamic structure of reality, inherent in how we structure and interpret the world, or both.

This all sounds grand, and it is grand. But we warn that all this talk of new beginnings, rebirth and triumphant marching from one enlightenment to the next, is but part of the story. We are perhaps still barely awake and, like the adolescent child, lacking in the wisdom and maturity necessary to understand and respect the pitfalls of hubris. We are not gods among beasts but merely a conscious organ embedded in the biosphere that is our womb. In turn, the biosphere is embedded in a vast cosmos, with as yet mysterious organization and awe-inspiring dynamics which ultimately inspire divine metaphors among gurus and scientists alike. Thought and waking consciousness give us freedom and power to begin to 'know' our cosmic origins, but as the sages warn, with freedom comes responsibility. If psychology has taught us anything about our origins and destiny it is that self-consciousness and mature self-realization is a temporary condition. We emerge into the world in a 'simple' and unconscious state and may depart in much the same condition. Succinctly put, the respective experiences before, during and after existence are:

nowhere — now here — nowhere

This view cannot be empirically confirmed beyond a certain point, and indeed there is much evidence to suggest that self-realized persons depart earthly existence in a state of advanced consciousness. It nevertheless raises significant questions about the unconscious and how we somewhat vainly try to understand it from our vantage point of consciousness.

Since the evolution of consciousness is so dynamic and relevant to almost any past, present or future endeavour, anything we say about it is likely to be a work in progress.[6] We have no absolute or standard vocabulary designed to please one scholar or another. We may veer more towards pleasing those of a holistic bent than those of an analytical persuasion. But our brand of holism is not quite the same as another's and is surely not some universal panacea designed to complete our incomplete understanding of the mysteries of existence. A full holistic explanation was never our intent. Nevertheless, it is our considered persuasion that the subject of consciousness must be tackled as broadly as possible. Since consciousness is for everyone, we hope the reader will find something for everyone in this book.

Some of our favourite gurus speak compellingly of 'integral consciousness'. Partly for this reason we stress integration of a broad range of disciplines from neuroscience and child development to evolution, anthropology, cultural philosophy and linguistics. One cannot discuss any of these topics without language, or more precisely without a multiplicity of languages and specialized vocabularies. This position is our justification for claiming that in an integral world two or more people can discuss or describe the same phenomenon with different languages and still be right. This does not mean that one may not also be wrong in one's assessment. Many people may describe a phenomenon in the same way and still all be seriously mistaken — as history has demonstrated on many occasions.

On this note we once again emphasize holism, integration and the multiplicity of 'meaning' that can be extracted from the phenomena we study. For some in the neurosciences, consciousness has come to mean human self-consciousness. (It is after all hard to study the states of non-self-conscious animals that have no referential language.) For us however, the meaning of consciousness is much broader and includes not only 'waking' self-consciousness, but also the subconscious and unconscious probed by early psychological pioneers such as Freud and Jung. We also assume that 'consciousness' is not located in any single organ or assemblage of neurons in the brain any more than the meaning of the word is located in the letter C, N or S or in an analysis of the chemical composition of the ink used in writing the word. Studies of how we perceive, how parts of the brain correlate with specific senses or components of senses — the so-called 'neural correlates of consciousness' (NCC) — are valuable pieces of the puzzle, but, in the same way that a protein or gene does not give us a complete understanding of an organism, so neurons cannot fully explain consciousness. We are after bigger fish, what we might call the Cultural Correlates of Consciousness (CCCs). What do cathedrals, poetry or quantum physics tell us about consciousness? What is it that makes us human? — our physical anatomy, our language, the position we allot ourselves on the evolutionary tree we have created, or is it perhaps, simply put, our consciousness, which embraces and defines all these factors and many more?

So much for our introduction to themes and subject matter: what about the authors? Where are we coming from? In keeping with our goal of integration, this book has been a collaboration between young and old, and between east and west. The book grew organically, and somewhat haphazardly, out of a course, taught for a decade by the 'senior' author, on the evolution of consciousness. The senior author teaches in America but is a British baby boomer with an academic background in paleontology. The 'junior' author is Japanese, from 'generation X' and has a background in psychology. So without further ado we invite the reader to sample our east-west, young-old brand of paleontological psychology.

PART 1.

Hello, Consciousness

1. Some User-friendly Notions of Consciousness

In the course of this book we will argue that modern biology and psychology are unable to separate the concept of mind (consciousness) from the body (biological development). We believe science is already proving that the body-mind is not just an integrated whole, but an integrated, and highly dynamic, evolutionary process. This conclusion is consistent with many perennial wisdom traditions that speak of an intelligent or conscious universe: it also appears to be the conclusion towards which holistic medicine, quantum physics/cosmology and many consciousness studies are pointing. However, the notion that consciousness is universal does not prevent us from perceiving a 'self' embedded in a world of many 'other' things. Although we begin with the polite formality of introducing the self to the others, we aspire to show that the consciousness adventure reveals a cosmos of infinite and endlessly intriguing complexity. This cosmos, in which our minds and bodies are so deeply embedded, has spawned countless creations including the thousand philosophies and adventures that give shades of meaning to our ever-changing inner and outer experience. Somehow our sentience and that of a billion cosmic cohabitants is enriched by mysterious sources of universal organization that both maintain the familiar structures by which we navigate and create the novel entities and experiences so essential for our physical and psychic nourishment. Although beyond full comprehension, we intuit that the illumination of our consciousness resonates with a greater light of universal spirit and that it is our destiny to explore the relationship.

You and me: the fundamentals of subject-object dualism

On the most fundamental level, the contemporary human experience of 'self and other' is often inescapable and enigmatic. It gives rise to the distinction between subject and object, subjectivity and objectivity creating so-called 'dualism'. The fact that this topic is widely debated in academic and philosophical circles perhaps proves the present importance of the point. Put another way, our fixation on these and other such themes tells us something about our present state of consciousness. Abstract debate aside, the concrete fact is that a 'dualistic' separation occurs in at least three crucial and early phases of our development as human beings:

1. At the time of conception, a new physical life (existence) emerges from the universal womb. A human being emerges from where, before, there had been nothing other than the conscious or not so conscious intent of the parents and their microscopic sex cells. From 'nowhere' we are 'now here'. An early phase of both physical and biological separation has manifested itself from some mysterious universal source; and moreover, the impulse to physically create a new being often begins with an idea!

2. Then, at the moment of birth, the individual is physically separated from its mother, and so the journey of individualization begins. This is a secondary phase of physical and biological separation.

3. This physical separation is not, at first, a full separation of consciousness. But usually by the end of the second year there is a momentous shift in consciousness when the child first recognizes itself as a separate individual or entity and utters the classic statement 'I'm me!' or some such articulation of the profound realization of the difference between self and other. This, then, is the critical first phase of psychological separation. This dualism manifests in almost all subsequent walks of life from intellectual debates, to party politics and

nationalism. 'I'm a liberal. I'm Welsh. I'm right.' So in the stark sense articulated by the philosopher Thomas Hobbes: 'Each of us is always and finally isolated from every other individual'.[1]

But wait! Before we go any further we should check ourselves and note that there can be no concept of separation without a concept of something unified from which separation occurs. Throughout the entire book we shall have to check how we represent the world to ourselves, including various evolutionary consciousness theories, and ask what this tells us about the nature of consciousness. In short, it is dangerous to commit to any hypothesis or model, without carefully exploring its origins and implications.

For example, it is important to realize that the physical separation (stages 1 and 2) does not constitute the psychological separation of stage 3. Thus, we could say that the observer, the parent for instance, might objectively view the child as a separate individual in stages 1 and 2, as abortion debates and attendant ethical considerations clearly show us. But, this perception has no meaning or reality for the child (the observed) until it becomes subjectively aware of itself. It is only at this moment of psychological independence that the differentiation of self and other, or subject and object, begins to mean anything to the child. (In strict academic terms one might say physical and biological separation precede psychological separation. It is also partly for this reason that there is ambiguity and debate about when the individual mind, psyche or personality is born.) We might also note that the change or 'shift' seems to be tied to an inherent stage of development — what we might call a biological phase shift. We might also say that the child is at once separate and not separate from the parent. This is an apt metaphor for life since to varying degrees we are all both separate and not separate from the world.

It is important to note that these three fundamental stages in the differentiation of self from other, which we all experience, follow an evolutionary sequence in time, or what we likely perceive as linear time. Later we will have to move beyond simple linear time models, but for now they will suffice as a starting point.

Equally important is the fact that this phenomenon of separation or differentiation has been much debated and is central to many perennial philosophical debates and classic psychological studies. For example, without the experience of a sense of separation, no debate about dualism would be possible. Such debates often concern abstract and scientific concepts peculiar to particular states of consciousness, and modes of argument such as the restricted convention of replying to courtroom questions only with 'yes' or 'no' answers. Indeed, as we shall later see, dualism cannot exist conceptually in certain states of consciousness. Nevertheless, dualism is meaningful in some contexts and must be honoured as a compelling experience — one we can explore and learn from.

Setting boundaries and getting centred

Based on the arguments just given, our subject-object consciousness necessarily perceives a 'boundary' between self and other. Ken Wilber makes the observation that one of the manifestations of self-awareness is the loss or diminished sense of connectedness/oneness with the cosmos. Thus, self-awareness is a shift that consciousness makes to alter its centre. Put another way consciousness of ourselves manifests as a type of self-centredness, or movement 'inward' that allows for an inner life. Again, during individual growth the development of a healthy, centred sense of self is considered both normal and desirable. On this theme it helps to note that there is a type of inherent compensation between the outer and inner worlds. Thus, as the individual grows physically, and expands his/her outer horizons or boundaries, so the inner space is enlarged and the centredness becomes stronger and more focussed. We are already on a road towards an awareness of the dynamic compensation between the inner, subjective and outer, objective world.[2]

Already, about 2,500 years ago the Greek Xenophanes was aware of the anthropocentric characteristics of human consciousness. He rejected Hesiod's idea of anthropomorphic gods like Zeus and so on, by claiming that if animals were capable of drawing, 'each would make the god's bodies have the same shape as they themselves had'.[3]

This same idea was humorously represented two millennia later by the Argentinean author Marco Denevi in his story *God of the Flies*, where these familiar insects conceptualize their deity as a giant fly. The assumption that animals and flies are self-conscious like humans might seem silly, or at best dubious, from our anthropocentric viewpoint, but the deeper truth is that we can project our self-conscious awareness to other species and imagine them creating gods just as we have done (even if they would never do such a thing!). Thus, Xenophanes' insight is that consciousness is capable of projecting (or transcending) itself beyond the physical realm. Leibniz (1646–1716) called these self-centred individual consciousnesses 'monads' whose fundamental characteristic is perspectival. They are the source of numerous self-centred ideologies and 'isms' — representations of the world that we create and then believe in as 'idols'. 'I'm a liberal. I'm Japanese. I'm right.'

The dualism of self and other indicates that our consciousness is the product of the constant amalgamation of inner and outer experience, since it must go outside of itself and come back in order to experience anything. This polarity between outer sensation and inner perspective formed the basis of most early research in psychology.[4] This implies that there is as much reality in our inner experience as in the outer one. In the case of the infant, this is self-evident because prior to the emergence of self-awareness there is no differentiation between the inner and outer reality: they are a fundamental unity. Indeed, philosophers (for instance, Plato, Locke, Berkeley, Hume, Kant, and so on) have often busied themselves debating over which experience is more 'real', and some have even claimed that our outer perception often seems to fail us (Plato) and/or fool us (Descartes).[5] However, we stress that no matter what viewpoint (epistemology) one holds, our inner experience (reflection), which is apparently lacking in most other species, has an impact on how we perceive outside phenomena and *vice versa*. The perennial pursuit to demystify consciousness (or 'know thyself') seems to be one of the defining characteristics of human self-consciousness.

Famously, Descartes denied the reality of his physicality, but nevertheless identified his mind as being embodied with tangible matter while declaring: *'I* think, therefore *I* exist'. (Meditation II) Here, it is apparent that Descartes' non-material mind could not have been able to identify itself as 'I' if he was truly non-tangible and thus lacked any boundary (or body). So our self-consciousness is self-identifiable so far as it has its centre, and its experience is mostly identified with its physical body. But, as noted, this is not always the case, and it is precisely the peculiar transcendental ability of consciousness to go beyond the tangible body that so intrigues the rational mind, and leads to the so-called mind-body problem that has the fancy historical label of 'Cartesian dualism'.[6]

The contemporary cognitive sciences and philosophy of mind have come to the conclusion that mind-body dualism is essentially wrong.[7] Consequently our dualistic mindsets have created many problems in understanding and 'explaining' our consciousness. Increasingly contemporary scientific studies suggest that we must recognize the artificial nature of many dualisms though some are inherent or inevitable products of our ever-changing consciousness. We agree with Diotima's reply to Socrates, who asked, if Eros is not beautiful and good, is it therefore ugly and bad? In reply, she said, ' ... do you really think that [if] a thing isn't beautiful, it's therefore bound to be ugly?'[8] Today we might speak of multiple perspectives, a plurality of worldviews or beauty being in the eye of the beholder. By the same token, the empirical irrationality of Cartesian dualism does not automatically point to its complete breakdown, since one of the primary characteristics of self-consciousness is to experience binary realities. Here the point is that the Cartesian dualism is neither right nor wrong but a consciousness structure, that can be viewed as a precursor to what later emerges in our evolving consciousness. So mind-body, or self-other dualisms are 'mere' manifestations of our self-consciousness expressing its separating and polarizing nature.

Our long-standing effort in understanding the nature of consciousness/mind has revealed that it defies most of the physical laws we have formulated; our consciousness is not only aware of its own activity here and now, but it is also capable of reflecting on the past and wondering about the future. Indeed these processes can be carried out simultaneously without much effort. This mental power, which ranges from the macrocosm *(kosmos)* to the microcosm (the quantum realm),

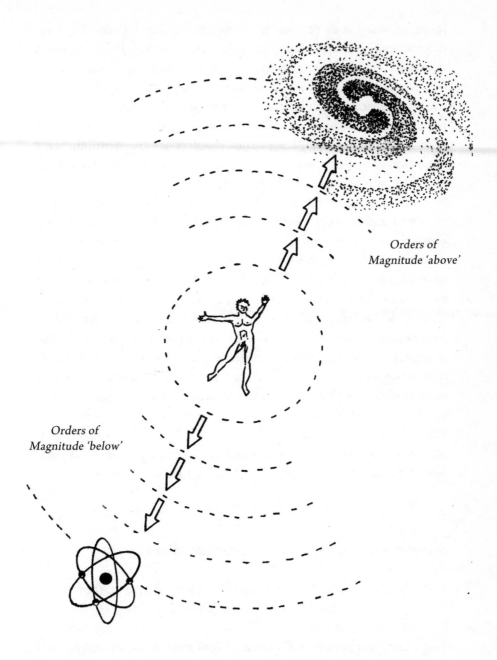

Orders of
Magnitude 'above'

Orders of
Magnitude 'below'

Figure 1.1. The human perceives many orders of magnitude above and below the scale of the body. Our perception places us in the middle of this hierarchy of scale.

seems to place us in the centre of the cosmos (see Figure 1.1).[9] This central position seems to remain 'relatively' constant. In other words the universe (literally 'the one thing') seems to expand in concert with our own sense of expanding consciousness: as in the microcosm, so in the macrocosm. Thus, we are at the centre of our external perception in most states of consciousness, with access to internal perception in various other states. As we shall see in later chapters dealing with biosphere organization and the anthropic principle, many scientific arguments also place us more or less at the perceptual centre of the universe. Thus, it is not surprising that there have been many heated debates over whether humans occupy a special, central place in the universe (or not). From our 'point of view' which is, by definition, anthropocentric, the answer is clearly yes! Although the same could be said for flies, we have no evidence that they have 'fly self-consciousness' or any inclination to consider such things. The very idea of a home, refuge or sacred space is deeply ingrained in our spiritual consciousness as stressed by Mircea Eliade and Carl Jung. These authors stated that we need these indispensable coordinates or reference points to centre or separate our consciousness in safe, sacred space away from the 'profane' chaos of otherness, out there beyond our control and comprehension. But alternatively, our ability to project a non-anthropocentric perspective, away from the sacred centre, and imagine what it may be like to be a fly, or to have another very different point of view, may also be a peculiarly human manifestation of expanding consciousness.[10]

Mind-body, mind-matter and other dualisms

'What is matter? Never mind! What is mind? No matter!'
GEORGE BERKELEY (1685–1753)

From a rational perspective, the paradox of the transcendental mind is that its focal point 'mostly' seems to remain embodied in our physical body, which we perceive affixed to certain time-space coordinates. Put differently, our conscious experience appears to be linked to the physical body, even though one can project one's thoughts and consciousness so as to imagine being somewhere else. One can imagine having a different experience, while still consciously 'assuming' that the source of these 'thoughts' is the physical body — in what is experienced as the body's present location at any given time. Nevertheless, it is important to note that, however proud or possessive we are of our thoughts, we cannot really control them too well. They pop into our head 'willy-nilly'. Our 'natural' identity with the body, which evidently requires thought or cognition, has not always been the case in human experience, as the infant stage shows; nor was it always the case in human history. There are many compelling reports of out-of-body experiences in which a person perceives their centre of consciousness to be 'completely' external to the body.

Teilhard de Chardin once stated that: 'We had thought that we were human beings making a spiritual journey; it may be truer to say that we are spiritual beings making a human journey'.[11] By the same token we might say that it is not clear if we are physical bodies having conscious experience or conscious bodies having physical experience. This same paradox is echoed in the quote of Chuang Tzu who said after falling asleep: 'I do not know whether I was then a man dreaming I was a butterfly, or whether I am now a butterfly dreaming I am a man'.[12]

As we hope to show, the human phenomenon is indeed an integral 'mind-body' or 'body-mind' experience much as Chuang Tzu and Teilhard de Chardin expressed.[13] But this is not mere poetry, though poetic expression may capture the experiential reality very well. We will in fact show that the mind and body develop as an integrated whole or 'system'. This is already well understood in the field of psychology where integrated physical, emotional and mental (cognitive) development is considered normal and healthy. But, we will take the argument a little further and show how consciousness actually moves like a dynamic wave, rising and descending in the body in a process wholly synchronous with biological development, and the ongoing physical and biological rhythms of the biosphere and cosmos. This process occurs on many developmental or 'ontogenetic' levels in the individual, from molecular and cellular to the whole body system, but it also reiterates or recapitulates in a fractal or recursive manner as an evolutionary trend in our species, our hominid family, and indeed throughout the biosphere. It is as if the organic system is a vibratory process of energy frequencies or waves that manifest physical form. The change in brainwave frequency as the body matures is just one striking and obvious example. Such observations bring us back to our initial position. Modern biology and psychology cannot separate the mind (consciousness) from the body (biological development) for they are deeply integrated. This integration is not merely a case of physical bodies being imbued with functional energy. Rather the flux of biological and psychological development is a dynamic, evolutionary 'process' not only in the individual, but also at the level of the species. Such conclusions encourage us to believe the ancient wisdom that the known and unknown are structured by a universal intelligence or consciousness which consistently proves one step ahead of our most intelligent insights. Put another way, our intelligent observations repeatedly discover that the biosphere and cosmos are a miraculous web of dynamic organization that must have been established long before we recognized it and attempted to describe it.

Observer and observed: a general perspective

As we proceed, we shall use the term 'perspective' in many different ways, just as we do in everyday language. For example, here are three different meanings:

— our 'general perspective' or view of the world around us (as discussed by Rudolf Steiner below);
— a peculiarly 'mental' and measurable view of the world in the sense used by Jean Gebser;[14]
— a three-dimensional view of space in the sense used by architects or artists.

For the time being, let us discuss only general perspective. If each consciousness has its own centre, boundary and experience, there must be multiple realities in which no absolute reality could exist. This might suggest that everybody's view is different and 'relative'. Rudolf Steiner suggests otherwise: he argues that a photograph of a tree is a two-dimensional representation of a three-dimensional entity, while the totality of the tree — the reality of the tree — can only be captured by multiple perspectives.[15] Obviously, the two-dimensional representation and the real tree are not equally valid, but, by the same token no representation is completely invalid. Thus, he emphasized that neither perspective is mistaken: each captures something about the tree, even though each representation is incomplete. In other words, Steiner points out that each entity, centre of consciousness, or monad (in the sense of Leibniz) is bound to perceive/experience a phenomenon from its own idiosyncratic space-time perspective.

Owen Barfield, one of the few early twentieth century thinkers to explicitly discuss the 'evolution of consciousness', took these ideas further when he discussed how we, as individuals and collective cultures, develop differing representations of the world, and too often mistake or 'idolize' them for some type of absolute external reality.[16] Barfield, an aficionado of Steiner, suggested that developing multiple perspectives was a helpful exercise in developing our consciousness and imagination. He spoke of different modes of alpha and beta thinking and what he called double vision: not the type induced by one too many beers, but the type instinctively recognized when we shift our attention to seeing the wood as well as the trees.

Understanding of such differences in viewpoint or perspective — or what we might call the changing dynamics of consciousness — helps us to embrace a more integrated and open-minded approach to the field. So we will disappoint anyone who has a one-model-fits-all view of consciousness. Integral consciousness has become something of a buzzword in current consciousness paradigms.[17] In turn, such integral consciousness insights lead to more complex cultural manifestations in our art, science and society. So our boundaries are ever shifting as the centre of our consciousness seeks more integral horizons. Like Aristarchus and Copernicus, we can shift the conceptual centre of the universe from the Earth to the sun, and beyond.[18] How different our consciousness becomes when its centre shifts from our Ego-bound bodies to the whole Earth, galaxy or boundless universe.

Some would suggest that this is the inevitable trajectory of evolution. We tend to agree, and moreover, have evidence to argue the case. In the chapters that follow we hope to do more than merely outline the history of consciousness studies, and lay out different academic theories. Rather we aim to show, in concrete terms, how the rational consciousness of modern western science, with its most recent foundations in the so-called 'Enlightenment' of the seventeenth and eighteenth centuries, is just one stage in the evolution of a more holistic and integral consciousness. In presenting and explaining the cogency of new consciousness paradigms, and the science and intuitive genius that supports them, we are demonstrating the very ongoing shift in intellectual and scientific consciousness that we consider inevitable.

In his study of *The Development of Autobiography in Western Culture*, Peter Abbs demonstrates just how recent and radical shifts in consciousness have been in western society.[19] According to the Oxford English Dictionary, it was not until 1674 that the word 'self' took on its modern meaning of 'a permanent subject of successive and varying states of consciousness'. The first appearances of compound 'self' words like self-sufficient (1598), self-knowledge (1613), self-made (1615), self-seeker (1632), selfish (1640), self-examination (1647), selfhood (1649), self-interest (1649), self-knowing (1667), self-deception (1677), self-determination (1683), self-conscious (1687) are all remarkably recent. Abbs' approach is reminiscent of Owen Barfield's classic work *History in English Words*, which used literature as an effective analytical tool for teasing apart the elusive sense of self-awareness on which so much of modern thought is now based.

As we embark on our evolutionary journey we hope we can recognize and integrate any and all of our more enduring cultural creations as equally valid expressions of the human consciousness experience. Ultimately, the nature of integral consciousness transcends linear time and linear models of progress, penetrating the fabric of evolutionary space-time to illuminate all fundamental manifestations of human cultural history, giving equal weight to art, science and spirituality.

Consciousness can do this. It may be the very fabric of our universe as we currently represent it to ourselves, and have done since the first sages in history pronounced on the subject. Consciousness is somehow ever-present in human being and becoming: it embraces, penetrates and creates a thousand languages, a million thoughts and theories, and a billion emotions, intuitions, impulses and imaginations, and a few trillion other tangibles and intangibles that are at once infinitely diverse and fundamentally united. All this and we have yet to turn to Chapter 2.

2. Through the Darwinian Paradigm and Beyond

Evolution evolving: how all paradigms change

Albert Einstein once said that: 'We cannot solve problems by using the same thinking we used when we created them.'[1] Likewise Ken Wilber expressed a similar sentiment in suggesting that scientists have to change their own consciousness before they can advance new scientific paradigms. On some levels, such thinking is part of our familiar everyday experience as we learn mathematics or foreign languages in school, gain on-the-job experience or learn to solve problems that we could not have tackled previously. However, researchers like Thomas Kuhn, who made an academic career out of studying so-called 'paradigm shifts' tell us that such shifts are not always easily accomplished because inertial forces, like conservative habits, are hard to overcome. Indeed there are those who make academic careers out of maintaining the status quo, usually out of genuine if misguided belief in the sanctity of their theories, but sometimes, regrettably, to obstruct innovation or thwart rivals. We are often over-conditioned into accepting old ideas and far too reluctant to adopt new ones. As we shall soon see, George Kühlewind and others talk about the way in which our reality is constructed through language. Thus, if our speaking becomes fixed, our view of the outer world also becomes fixed. This was again the perspective of Owen Barfield who, in his *History in English Words*, showed how countless words, like 'mechanism' for example, become fixed representations of how we believe the world 'works'. In this sense, old ideas are merely the lenses through which we see the world. Likewise Gestalt psychology suggests that new learning occurs not by trial-and-error (a quantitative shift) but by insight (that is, a qualitative shift).

We stress that we are not talking about making odious comparisons between the old scientific ideas from previous generations that are probably wrong and new ideas that are likely to be right and so 'better'. This is what C.S. Lewis called 'chronological chauvinism', or what today we might call age discrimination. There are plenty of examples of old ideas that were well ahead of their time but were nevertheless rejected by persons who claimed to be up to date when they were in fact 'behind the times' in their thinking. Famous examples include Wegener's theory of continental drift, and the work of the Greek Aristarchus who, almost two millennia before Copernicus, held that the sun was the centre of the solar system (see Chapter 1). Thus, chronological time is not a reliable criterion for judging the validity of an idea. ('Love thy neighbour' is clearly a better old idea than twenty-first century war.) Indeed, as already noted, the new physics seems to be in the process of discovering what some ancient wisdom traditions already knew about energy fields and space-time integration. The only difference was that these ancient traditions had not expressed their observations in mathematical equations or by using causal or mechanistic explanations. So, what we are really talking about is shifts in consciousness that allow new improved paradigms to come to light.

One such paradigm is what is often labelled the Darwinian Theory of Evolution. As recent debates show, even at the dawn of the twenty-first century this theory is not universally accepted, at least not in all details, even by many within the scientific community. We shall explain why this is so, and discuss the strengths and weaknesses of various gross and subtle versions of evolutionary theory. But before we do this we should stress that this is certainly not a simplistic debate between scientists and creationists, as to whether evolution actually takes place or not. The issues are far more complex, involving scientists, philosophers and thinkers of many persuasions.

Central to our purpose to introduce a new perspective on the evolution of consciousness, is a review of the whole idea of evolution. The great strength of the evolutionary concept is that it speaks to the changing and dynamic nature of organic systems or what we might call organic reality. Thus, we can talk of the evolution of plants, animals, culture, cosmos and consciousness, not to mention the evolution of everything from kitchen utensils, to phones and automobiles. Indeed, there is a burgeoning interdisciplinary literature on this very topic: 'the evolution of everything'. This in turn has led some to the idea of a grand evolutionary synthesis which brings about the integration of biology, consciousness and social systems.[2] Surprisingly, we shall find that the famous paradigm of Darwinian evolution lacks full support in some quite sophisticated scientific circles. Perhaps, to put a more charitable spin on the matter, we could say that the theory of evolution is evolving. What is often in question is not whether natural selection is a useful paradigm for competitive biological systems, but rather: does it work as an explanation for complex, ecological, sociological and psychological systems? We will use the famous theory of Charles Darwin and Alfred Russel Wallace to show exactly how this evolutionary change (in evolutionary theory) is coming about, and what type of new scientific consciousness is replacing it.[3] In the vocabulary of many current pioneers we can talk about an evolutionary shift from the mental to the integral consciousness paradigm.[4]

Darwinism, mechanism, materialism — and their shadows

The ancients tended to view the cosmos or nature organically. Humans lived close to, not separate from, nature. This lack of separation was both a psychological state and a technological reality. Until the Renaissance and the Enlightenment, modern western science had invented few machines and language had adopted few machine metaphors.

The early biologists like Johann Wolfgang von Goethe and Geoffroy Saint-Hilaire developed very holistic and organic ways of thinking, which are still useful today.[5] They used terms like 'transmutation' and 'metamorphosis' long before the terms 'evolution', and 'mutation' were widely used. In short, these terms signified dynamic change. For many with only a smattering of biological education, the term 'metamorphosis' conjures up an image of a butterfly emerging from its chrysalis, or some mysterious transformation of rocks or psyche. So transformation can and does occur on the physical, biological or psychological level.

Darwinism had many precursors, but because they are too many to mention, they are usually reduced to a few rather misleading stereotypes. The most simplistic is that the pervasive pre-Darwinian belief in a creator deity led people to believe that each species was created separately, and humans, blessed with the gift of rationality, had a special and 'separate' place in nature. In this view, this place was ultimately at the top or culmination of what Plato and Aristotle regarded as a great chain of being — itself an early evolutionary concept.

However, some leading biologists of the late eighteenth and early nineteenth centuries, especially the famous George Cuvier, firmly believed in the fixity of species. Cuvier's view was static or atomistic rather than dynamic like that of Geoffroy Saint-Hilaire. Goethe, with typical insight, noted that, ultimately, it was hard not to vacillate between the two poles of atomism and dynamism, or between matter and process. In short, things change, but only so much: they continue to maintain their characteristic form to a remarkable degree. This is one of the simplest and most compelling of organic philosophies, attributed to Douglas Fawcett.[6] Conservative forces maintain our form, and prevent our disintegration, while novel forces induce change and prevent us from complete stagnation. The two forces exist in a necessary dynamic tension, but they create what appears to be a contradictory paradox. Thus, like life and death, electrical positive and negative, or light and dark, novel and conservative forces *must* coexist.

Darwinism, however, as it emerged in the mid-nineteenth century, was an interesting mix of atomism and dynamism. On the one hand, it was based on the general idea that animals underwent gradual transformation through long periods of geological time. Geology at this time was a new science and, as a manifestation of human scientific consciousness, it was just getting to grips with the concept of what we now call deep time. Talk of millions of years of Earth history was both novel and radical. On the other hand, despite Darwin's deep appreciation for geological time, his theory of natural selection was based on his observations of artificial selection in the farming, stock breeding and horticultural world. In addition he was influenced by the ideas of Thomas Malthus (1766–1834) regarding the so-called 'population problem' which held that populations could not increase without generating fierce competition for scarce resources. This view tended to reduce organisms to atoms, or individual material entities, forced to compete with one another in a process that the British philosopher Herbert Spencer called the 'survival of the fittest'. This, as we shall see, emphasizes the competition model at the expense of the cooperation paradigm.

Both Darwin and Alfred Russel Wallace, co-discoverer of the theory of 'evolution by natural selection', were products of the Victorian era. Despite being very different personalities, they both were inclined towards seeing geological and biological evolution as a slow and gradual process that mirrored the orderly, stately development of science at the time. Indeed their theory of gradual change, later known as 'gradualism', germinated quite gradually when first introduced, and so was barely noticed by most of their contemporaries in 1858, and certainly not regarded as radical, even by their closest colleagues. It was only with the publication of *The Origin of Species* in 1859, that the world suddenly took notice, as the implications that perhaps humans were descended from apes sank into the general consciousness. One of the theory's implications was that with the discovery of a 'mechanism', one could argue that no divine guidance or intervention was necessary for evolution to proceed. (Nevertheless, Darwin and his contemporaries understood that the origin of life, and the evolutionary process, was still mysterious and not obviously explained by selection.) When such mechanistic rationalizations were linked to Marxist ideology, Darwinism came to be identified by some with a shadowland of implied atheism that has muddied debate ever since. Although such a spin is but one possible inference, it has often been taken as the basis for branding evolution as anti-religious or religion as anti-evolutionary. Ultimately no such inference is necessary.

However, as noted below, Wallace, co-founder of the selfsame theory, had an entirely different and perhaps much healthier view. Darwin, who had had a formal theological education, was not alone in feeling the retreat of divine comfort, and many have remarked on the anguish he and his family suffered as his evolutionary theory promulgated a sense of 'separation' from Godly succour. Had twentieth century psychoanalysts like Erich Fromm been around to advise, they would have noted the powerful tensions set up by this sense of separation. Paradoxically, Wallace, who had no formal religious affiliations, was not conflicted. His was an overtly spiritual sensibility, uncomplicated by the influence of any institutionalized religious creed. Although Wallace has been wrongly labelled as a non-scientific 'spiritualist' he was simply a gifted renaissance man with a deep scientific interest in the psyche at a time when psychology was just emerging as a discipline. Many other scientists of the day shared his interest in attempting a serious 'scientific' exploration of the paranormal. Ironically, when it came to evolutionary theory, Wallace preferred Herbert Spencer's 'survival of the fittest' metaphor over Darwin's term 'natural selection' which had the connotation of divine intervention.

Although it is beyond the scope of this book to psychoanalyze Darwin or any of his like-minded contemporaries, there can be no doubt that Alfred Russel Wallace provides a fascinating psychological contrast, not just for his extraordinary personality, but for the manner in which he came by his evolutionary insights. While Darwin, an inherently cautious, shy and conservative man ruminated cautiously, even fretfully, for many years, in the seclusion of his quiet English home, Wallace received his insight into the natural selection process in a sudden flash during a feverish bout of malaria in the far-off Malay Archipelago. Likewise, while Darwin focussed his attention on details of biological evidence that would support or undermine the theory, and worried as to how it would be received, Wallace embraced the discovery process as if it were a perfectly natural part of a greater cosmic evolutionary process, in which we all participate. Not only did he generously encourage Darwin to take his full share, and more, of the credit, but he enthusiastically extrapolated his evolutionary thinking to almost every sphere of human endeavour. He wrote, lectured and debated, often with extraordinary insight, on botany, entomology, ornithology, biogeography, anthropology, linguistics, medicine, economics, political reform, the dangers of militarism, astronomy, the universe and spiritualism. It is for his interest in the spiritual nature of man that he is most often unjustly reprimanded by sceptical and conventional scientists. But, as the twenty-first century dawns, Wallace looks more and more like a truly holistic thinker who was well ahead of his time. It is this alternative 'Wallacian' consciousness that we refer to in our section heading as the 'shadow' that stalks the Darwinian paradigm, showing up some of its less subtle mechanistic and materialistic aspects. Wallace perhaps embodies a subtle shift in consciousness that allowed him to broaden his horizons and look holistically at evolution as a physical, biological and psychological process. As one biographer remarked, Darwin's last book was about worms; Wallace's was entitled *Man's Place in the Universe*.[7]

The conservative and the progressive are two species or modes of consciousness that need not be judged or compared unduly. They are complementary in a general sense, if not necessarily compatible with a given paradigm. In the context of our present discussion we may be a little dismissive of certain 'old' ideas that do not resonate with the 'new' ideas we wish to develop: but, as noted already, some of our ideas represent a re-evaluation of ancient wisdom, and we too wish to avoid the trap of 'chronological chauvinism'. If any one paradigm is to be favoured, it is that of holism, not because it is better than the alternative (reductionism) but because holism has been overlooked and now needs a fair hearing — if only for balance. It may also simply be that in this age of interdisciplinary globalization, consciousness is shifting in the holistic direction and that, sooner or later, we are all along for the ride.

Before we go in this direction, we can note how tenacious the Darwinian paradigm has become. This is a testimony to its importance but also to the power of conservative intellectual forces reluctant to give way to novel influences. Many standard new species of Darwinism have been spawned in the twentieth century. The first of these was labelled as neo-Darwinism or the New Evolutionary Synthesis (not to be confused with Laszlo's Grand Synthesis of Evolution). Its popularizers included Julian Huxley, grandson of Thomas Henry Huxley, Darwin's outspoken supporter remembered by history as 'Darwin's bulldog'. With the development of the science of genetics and the discovery of the role played by genes in the inheritance of acquired characteristics, the basic tenets of natural selection were applied to ever more mathematical, statistical and theoretical population studies (population genetics). While statistics swept the academic world, evolution came to be regarded as a probabilistic game of chance in which random mutations (unpredictable external influences) actively impacted hapless or helpless and rather too passive organisms. Thus, the powerful idea of natural selection was transferred to the molecular or gene level. In short the implication was that organisms were largely at the mercy of their genes. This idea reached its zenith in the notion of *The Selfish Gene*, promulgated by Richard Dawkins, who argued that genes use host organisms as a way to perpetuate their own lines of inheritance.[8]

The gene might also experience a random mutation (rather like an act of God) which would start a new line of inheritance to be tested in the cut-throat battle for survival. Hidden in this inference is a familiar theme, that the cosmos is a stable process interrupted periodically by unpredictable random events. The idea that we are susceptible to these random events is unsettling psychologically and gives rise to the idea that we may be condemned to passivity, and unable to act to control our own destiny. Indeed, it gives rise to a deep-seated fear about our individual and collective survival as a species. This is perhaps an inherent weakness in the Darwinian/neo-Darwinian paradigm. It assumes that we are vulnerable to our genes and our environment. It is not coincidental that this seems to place us between a rock and a hard place, between the unknown capriciousness of the genetic microcosm and the unpredictability of the universal macrocosm. It is indeed intriguing that we seek explanations at the extreme molecular and cosmic limits of our perceptual range, while all the while we abide in the centre as our own, but as yet poorly understood, organism. (See Chapter 1, Figure 1.)

This Darwinian paradigm has been projected into two other related fields. The first is Sociobiology, which Harvard biologist Edward Wilson also dubbed 'the new synthesis'. Sociobiology attempts to integrate biology and physiology with psychology and sociology. This theory essentially applies the tenets of Darwinism to the evolution of behaviour. The second field is Evolutionary Psychology, which is a further permutation of Darwinian and sociobiological paradigms that like to talk about how our thinking and behaviour is 'hard-wired' into our psyche as a result of behaviour patterns which we can trace back to the Ice Age and the very dawn of our species. Perhaps some more physical behaviours like eating and mating are long-established habits that were originally unconsciously motivated, but evolution in the emotional and mental realms is probably another matter. The idea of wired circuits repeating stereotyped behaviours is both materialistic, and computer derived, and significantly at odds with the whole idea that consciousness evolves dynamically, or the notion that we can exercise choice. As we shall see when we examine Jean Gebser's theories, consciousness is just as easily portrayed as dynamic, fluid and capable of rapid transformation. As Deepak Chopra recently suggested: 'the trend of time must obey consciousness when it decides to change'.[9]

Just as Richard Dawkins projected his idea of selfishness onto genes, so he and others have come up with the idea, reminiscent of Hobbes, that our ideas are in competition for survival. For linguistic symmetry these ideas have been called 'memes'. Ironically a meme is indeed selfish since it appears to cry 'me me!' In recent years the meme concept has come under increasing criticism for being a hopelessly 'self-referential' concept. Critics include the British philosopher Mary Midgley and the Oxford theologian-biophysicist Alistair McGrath, both of whom argue that the meme theory is blatantly unscientific for many reasons, not least of which is that its claims of objectivity are hopelessly subjective. As discussed later, these rancorous debates appear to tell us something about the polarizing forces that affect science and philosophy.[10]

So Darwinism has had a great influence on our thinking. The current literature on cognitive sciences and philosophy of mind still leave one with the impression that most writers take Darwinian evolution for granted. In other words, they see evolution from primates to humans as a logically necessary sequence for justifying the emergence of higher consciousness in the human species. In addition, the Darwinian model, which is essentially linear and economic in nature, mostly promotes the mental/rational win-lose, either-or manner of thinking. While we support the idea of organic evolution as a dynamic reality, Darwinism is no longer regarded as an adequate explanation for *all* aspects of evolution, especially those that have to do with psyche and consciousness. Thus, juxtaposing discussion of Darwinian evolution with other paradigms, especially Gebser's cultural model, Goethe's organic model (Chapters 5–6) and various eastern philosophies, as below, we can throw new light and focus on the differences between the Mental (reductionist-analytic) and the Integral (holistic-synthetic) ways of thinking. This differential approach is in itself a manifestation of the evolution of scientific thinking and a shift in the consciousness paradigm.

One of the most innovative evolutionary theories to emerge in the twentieth century was that of symbiogenesis or simply symbiotic evolution. It is perhaps no coincidence that this theory was developed by a woman, the biologist Lynn Margulis.[11] In its simplest form this theory is one that emphasizes cooperation rather than competition. So again it is a shadow, feminine or flip side of the masculine Darwinian coin. The primary tenet is the principle that one should not kill the goose that lays the golden egg. Organisms may need to feed on one another, but they cannot survive if they kill off their entire food source. The theory has many other more subtle and scientific aspects that help explain major evolutionary events such as the appearance of complex cellular life. The theory has proven that some micro-organisms represent cooperative symbioses of microbes that used to live separately but now are joined in a mutually beneficial marriage, dividing and sharing their labour. They cooperate, in flagrant violation of Hobbes' dictum that 'minds never meet'. This theory shares the language of altruism with sociobiology and raises interesting question about where to draw the line, if any, between what is good for the individual and what is good for the community. It shifts the emphasis from the win-lose paradigm of competition to the win-win paradigm of cooperation, also promoted by Elisabet Sahtouris, another female Ph.D. in biology.[12] So the Darwinian evolutionary paradigm is softened as it evolves from a 'nature red in tooth and claw' paradigm to one in which the world is potentially more benevolent and rewarding of cooperation.

Shifting to the next critical phase: beyond linearity

Darwin was puzzled by the fact that the fossil record did not always show evidence of slow gradual change from one species to the next. In fact, he worried that the mysterious and sudden appearance of new species was a serious threat to his theory. As a result, he never really demonstrated the phenomenon implicit in the title of his book, nor have any subsequent evolutionary scientists. The 'origin of species' remains a mystery. As we have seen, our intellectual consciousness, or what Jean Gebser calls the mental consciousness structure, is predisposed to the abstract activity of defining, measuring and explaining the world with various essentially dualistic scales: big-small, long-short, slow-fast, and so on. So the opposite of gradualism is suddenness.

Thus, it is not surprising that two paleontologists, Niles Eldredge and the late Stephen J. Gould, have attempted to resolve Darwin's dilemma by proposing the theory of Punctuated Equilibrium, which argues that the fossil record indeed shows long periods of very gradual change or near stasis, punctuated by sudden jumps forward: what we might call evolutionary 'quantum leaps'.[13] Again the so-called 'mechanisms' which produce such sudden novelty from the slow-changing, static or conservative state, are essentially unknown and much debated, but it is recognized that somehow the system reaches a 'critical point' which forces or facilitates change.

What is novel about this theory is that it is pluralistic. It accommodates both stasis and change: our conservative and novel forces. Although not explicitly discussed by Eldredge and Gould, the theory of Punctuated Equilibrium appears to resonate with the idea of a 'phase shift', in the sense used in physics, where a substance, like water, may change from one state to another: from solid to liquid or gas. This type of thinking is also resonant with the new science of chaos theory. From this viewpoint the world is seen as series of energetic systems that oscillate or vibrate at a particular frequency for a given period, and then shift or jump to a different state.[14] Chaos theory evolved as a synthesis of other modes of integral thought such as General Systems theory and Cybernetics. A central tenet is the idea of a chaotic attractor, a state or pattern of activity towards which a system tends to slide of its own accord. This may be a static attractor in the case of a pendulum that comes to a full stop. However, as suggested by Allan Combs, co-founder of the Society for Chaos Theory in Psychology, there are many other types of attractor that can be considered in the study of psychology and consciousness.[15] The human circadian rhythm of sleep and wakefulness approximates a fixed cycle attractor. Other cycles are more complex and referred to as chaotic or strange attractors and may include such complex phenomena as physiological and emotional cycles, mental states and so on. One of the characteristics of many of these attractor systems is that although they repeat in a cyclic pattern, the cycle is never quite the same, like the pen on the pendulum that traces a beautiful organic pattern (Figure 2.1). In technical terms this is called recursion.[16] It is not strict repetition, which is like photocopying, but rather it is like a habit or memory, which varies slightly with each cycle. Such 'patterning 'is very organic, like the cycle of our days and seasons, where the same old conservative pattern operates on one level, but always introduces something new on another. In recent years many people have become familiar with another appealing version of this type of systems organization known as fractals. (See Figure 2.2.) Here again we see the dynamic interplay between conservative and novel forces as a creative process and a thing of beauty.

The phase shift paradigm is very useful in thinking about shifts in states of consciousness. We have already seen that the infant undergoes a shift from simple to self-consciousness usually in the second year of life. This dawning of awareness of self is so fundamental that it really defines the emergence of the most essential psychological trait of being human. It is a developmental stage that is an inherent part of human development, though it seems to require the stimulus of social interaction with other humans and the normal development of language. (In rare cases where feral children are raised by animals, normal human self-consciousness and language do not develop. Instead the child adopts a consciousness resembling that of the animal foster parents.)

While self-consciousness lasts a lifetime, another shift in consciousness may occur in mid-life. This shift, which does not appear

Figure 2.1. A pendulum traces recursive patterns.

Figure 2.2. The fractal principle illustrated. (Modified after Mandelbrot 1997.)

to be universal, has been described in many ways. In 1901, the eminent American psychiatrist Richard Bucke referred to this shift as the Cosmic Consciousness experience,[17] much like a revelation, epiphany or the 'religious experience' described by his famous contemporary the psychologist William James.[18] These authors considered that those who experience such shifts are 'consciously' aware of an enhanced intellectual and moral 'sense' that permanently modifies previous states of consciousness towards a more 'sage' and universal awareness. As we shall see in Chapter 3, Bucke claimed the shift typically takes place in mid-life and is often marked by a pronounced increase in creative activity. These shifts probably correlate with recognizable phases of biological development (for instance, seven year cycles). Cosmic consciousness may be largely synonymous with mystical experience described in very similar language down through the millennia (see Chapters 8–9). Indeed it is an interesting psychological fact that most humans recognize something special about people who have had such enhanced consciousness experience, often regarding them as prophets or messengers from 'God'. Thus, twentieth century consciousness teacher-philosophers like Alan Watts, who continued the tradition, speak of cosmic consciousness in matter-of-fact terms, assuming it to be a consistent or at least desirable and attainable stage of mature human development.[19]

Thousands of 'sudden' religious epiphany experiences have been recorded and studied by scholars like the late Alister Hardy, a biologist who founded an institute for the study of religious experience at Oxford University (and now moved to Lampeter in the wilds of mystic Wales).[20] Such epiphanies may be manifestations of a type of psychological punctuated equilibrium; although the change may be gradual, it is often very abrupt. In recent years neuroscience has taken increased interest in the chemical, neurological-electrical (brainwave) and behavioural correlates of altered states of consciousness manifest by those adept at transcendental meditation.[21] The ability of such individuals to alter their states of consciousness, implies that such shifts can be identified, measured and self-induced to varying degrees.[22]

There are collective equivalents, too, of such phase shifts in consciousness. Peter Russell has suggested that as the Earth's population (currently between 6–7 billion) reaches critical levels of connectivity, a Global Brain phenomenon may be activated.[23] This idea, which was foreshadowed by the visionary paleontologist-priest Pierre Teilhard de Chardin in the 1930s, is popular with students of cybernetics and complex systems.[24] Teilhard labelled the growing network of global communications and thought as the 'noosphere', and regarded it as an evolutionary development that was as revolutionary as the origin of life. The evolution of a cosmos from one dominated by inorganic matter (physiosphere), to one witnessing the emergence of life (biosphere) and human consciousness (noosphere) can be seen as a series of momentous phase shifts. The Global Brain paradigm suggests that it takes about ten billion (10^{10}) atoms to make a complex organic molecule, 10 billion molecules to make a cell, and ten billion cells to make a complex multicellular organism. This raises the question as to what shift in global organization or consciousness might occur when the world population reaches about ten billion.

Another famous example of this critical-mass phenomenon is the so-called Hundredth Monkey Effect. This controversial idea originated with observations of macaque monkeys living on Japanese islands. After one group learned to wash sweet potatoes, a critical mass of knowledge seems to have been reached and the learned behaviour 'jumped' to monkeys on another island that had not had contact with the original group. Several well qualified British biologists have written on this subject including Lyall Watson and Rupert Sheldrake.[25] While some suggest the phenomenon is proven others dispute the evidence. Here we only need note that the idea is convergent with Global Brain, phase shift, critical mass and punctuational thinking inherent in cybernetics and complex systems paradigms. Such thinking has worked its way into our vocabulary in the use of such terms a 'quantum leap'.

Notions of transforming our individual or collective consciousness through positive thinking, intentionality, prayer vigils or peace marches all point to a belief in these subtle energies. Whether or not such positive thinking can actually etch new pathways into an otherwise conservative brain, as is often claimed, there can be little doubt that our evolving consciousness places increased confidence in the power of the mind to change the world.[26]

Perennial philosophies of holism

Movies like *What the Bleep do we know* have popularized the idea that we live in a quantum universe where energy, matter and mind are complexly interconnected. We can even begin to make our dreams, or at least our thoughts, come true simply by exercising the power of intention.[27] Selling such ideas has become lucrative in many so-called 'New Age' circles, but perhaps the important question here is: what kind of shift in consciousness is indicated by such perspectives? So what is the quantum universe? Quantum Physics developed between 1900 and 1930 as a marriage of general and special theories of relativity developed by Ernst Plank, Albert Einstein, Niels Bohr, Wolfgang Pauli, Edwin Shrödinger, Werner Heisenberg, and others. In the quantum universe the concept of linear time breaks down. This paradigm shift began with Einstein's observations that with acceleration to the speed of light, time slows to a virtual standstill, while mass increases towards infinity. These and other related concepts demonstrated the strange properties of matter. As a result, we can no longer be certain we can measure all the properties of particles like electrons at any given time. Post-Einstein quantum physics shows that when one impacts or 'observes' a particle in one location there can be an instantaneous effect on another particle at another location. This suggests matter, energy and consciousness are nonlocal.[28] As discussed in Chapter 7, these results blur the distinction between the observer and the observed, and create profound scientific and philosophical ambiguity when we try to speak of objectivity and subjectivity.

Although the mathematic expression of this work is complex and difficult to understand, during the twentieth century the concepts have filtered down to general consciousness, informing us of the mystery of the universe and the ambiguity, insubstantiality and quirkiness of matter at the subatomic level. Modern physics has recently focussed on seeking theoretical models that tie together all we know about the physical universe: that is, the laws affecting gravitation, electromagnetism, strong and weak nuclear forces, and so on. These theories have labels like GUTs and TOEs referring to Grand Unified Theories and Theories of Everything.[29]

Ervin Laszlo, author of more than eighty-three books, has demonstrated a profound interest in the integration of scientific theories. In *Evolution: the Grand Synthesis*[30] he shows that one can view the hierarchy of physical, biological and psychological (or mind) realms as a hierarchy of increasing complexity that correlates with decreasing physical strength of bonding (see Figure 2.3). At the fundamental, atomic level of organization we encounter only a few elementary particles held together by extremely powerful nuclear and magnetic forces. These comprise about one hundred mostly stable elements of the periodic table and are the building blocks from which other compounds are constructed as molecules. Such compounds are far more numerous, running into many thousands of mineral and organic species ranging from quartz and simple salt to complex proteins and enzymes. But unlike the hard-to-split atom, chemical compounds are relatively easy to separate and recombine as we learn in our introductory chemistry classes. Molecular bonds are far weaker than their atomic counterparts and can be broken by such simple external force as heat.

As one moves up the periodic table of elements from Hydrogen (atomic number 1) to Calcium (20) and elements like Uranium (92), so the elements become more complex, with multiple layers or 'shells' of electrons. Likewise, while some compounds like water (H_2O) or salt ($NaCl$) are relatively simple, others, especially the organic species, are very complex. Proteins for example, of which humans have about 30,000 distinct types, are long chain molecules. The well known example of hemoglobin has the following formula: $C_{2952}H_{4664}N_{812}O_{832}S_8Fe_4$ which, technically, adds up to some 9,272 individual atoms! The relationship between organic molecules is highly dynamic and 'loose' in comparison with the simpler compounds discussed above. So a process like the Krebs cycle, also known as the citric acid cycle, involves a cascade of biochemical changes taking place in a complex milieu of cells and tissues.

At the level of cells and tissues of living organisms, the number of atoms and molecules involved runs into billions. Living organisms have many different cell types (humans are made up of about 120 distinct

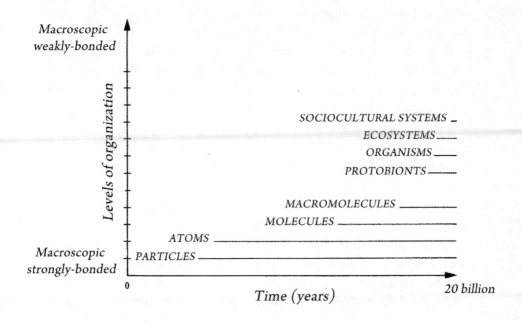

Figure 2.3. The size-organism bonding continuum. (After Laszlo 1987, Fig. 1.)

cell types) while different species of life on Earth are too numerous
to count. Estimates range from about 1.7 million described living
species to an estimated thirty million. But this does not include extinct
species, which may add up to hundreds of millions throughout Earth
history. At higher levels of organization, organisms defy description
in purely physical terms. It is hopeless to try and define the cohesive
forces that bind cells, tissues and organisms in terms of the strong
nuclear, electrical and chemical forces that bind simple elements and
compounds. As we shall see in Chapter 7, organisms have coherence
and field-like alignments of cells and tissues referred to as molecular
democracies with complex optical properties characteristic of liquid
crystals. The organism is a system in constant flux with gas and fluid
exchange between cells, tissues and major organs. Add to this the flux
of the body's electromagnetic field and it becomes apparent that we are
hard pressed to define the physical body in simple terms. The body then

is a dynamic flux of organization like an ecosystem. Taken collectively, ecosystems such as reefs are highly complex with many layers of interaction, coordination of breeding and maturation cycles and so on. In such a system the 'ties that bind', the modes of communication and interaction, whether obvious or subtle, are so complex as to defy classification in any simple or rigid scheme.

In 1979 James Lovelock introduced the Gaia Hypothesis.[31] The basic tenet of this hypothesis is that the Earth's biosphere is a single superorganism. Support for this holistic idea is that all the parts function together in a state of complex homeostasis. If the planet were dead, it would run down to a state of inert chemical equilibrium such as observed on Mars. By contrast, homeostasis refers to the complex systems of feedback in which plant and animal life sustain one another and the atmosphere and ocean composition, through complex circulation systems such as the carbon and sulphur cycles. Thus, the Earth's vegetation is like the planet's lungs, its fresh and salt water systems like the planet's arterial and venous circulation systems, and soils and seabed like the digestive system. The Gaia hypothesis is inherently holistic, stressing the interdependence that was first realized when ecology became a recognized scientific discipline in the 1960s. It is noteworthy that Lovelock chose an ancient symbol — Gaia, the ancient Earth Goddess of the Greeks — to symbolize his scientific hypothesis.

Lovelock developed his Gaia hypothesis through studies of atmospheric chemistry. However, the general message is that the Earth is our home, our mother. This worldview is user friendly to the non-scientist who can appreciate the message of ecology (*ekos-* actually derives from the Greek word for home). Thus, human consciousness has evolved to be able to appreciate our place in the biosphere (Gaia) and the whole universe both from the scientific and the symbolic perspective. Our theme here is the integral or unitary perspective: the goal to see the wood *and* the trees with a healthy double vision.

We can describe the complex biosphere and universe in terms of chemical and mathematical equations, or in terms of more general language of ecological interconnectedness, and even in terms of mysterious quantum and gravitational phenomena, density waves or

Akashic fields.[32] The various languages are complementary, but all are abstract 'representations'. $E=mc^2$ is not actual energy, mass or light any more than the words are the real thing — the phenomenon itself. The menu is not the meal. The map is not the territory. We constantly convert what we perceive as reality into representations or symbols, and we should remember that this process is itself the manifestation of a dynamic, evolving consciousness.

To illustrate this point we shall shortly turn to a concrete example of how early biologists like Goethe and Geoffroy Saint-Hilaire already anticipated the integral systems approach to organic organization. For example, the idea that all vertebrate organisms essentially manifest variations of an archetypal body plan conveys much the same idea as the notion of fractals or chaotic attractors. Biologists like Wolfgang Schad have developed Goethean biological thinking to a remarkable degree and have managed to show how organic systems are hierarchically organized along similar lines, and how they are influenced by different developmental 'time' scales.[33] Their approach, however, is not abstract, or mathematically representational as in the case of systems and chaos theorists. Rather it directly portrays organisms as manifestations of a higher level of organization that was previously hidden from our view simply because we were looking with analytic eyes, rather than with the synthetic, organic eyes of nature. As we already indicated, it is easy to mask a deeper organic reality with a mechanical representation or conceptual system of our own making. We will demonstrate this type of fundamental difference in perception or perspective quite clearly in a subsequent chapter. And we will show how complex biological systems can be appreciated much more deeply without the use of complex mathematics.

The term 'Perennial Philosophy' was coined by Leibniz, but popularized by Aldous Huxley, for whom it was both a primary interest and book title.[34] He referred to ' ... the one, divine Reality substantial to the manifold world of things and lives and minds', and suggested: 'the nature of this one Reality is such that it cannot be directly or immediately apprehended except by those who have chosen to fulfil certain conditions, making themselves loving, pure in heart, and pure in spirit'. Such suggestions of lower and higher 'levels' or states of consciousness recur as part of a perennial debate throughout the history of consciousness studies, and have to be addressed, as many authors such as Ken Wilber have attempted to do (see Chapter 1). Again a dualistic, mental element creeps in when trying to measure consciousness on any kind of scale, but it is also fundamental to human development as can be seen in many standard studies of cognitive (and moral) development.[35]

Hints of Perennial or neo-Perennial philosophy were present long before Leibniz and Huxley. In the West, the philosopher/mystic Plotinus (AD 204–270) formulated a spiritual cosmology, structured along the lines of the trinity, with three *hypostases*.[36] The term hypostases refers to underlying attributes, in this case: the One (the Divine), the Intelligence (Forms), and the Soul (Psyche). Similar hypostases can be observed in the system of the Indian guru Sri Aurobindo, as noted in the works of Ken Wilber.[37] Plotinus believed that all existence emanates from the dynamic unity of these three levels, and the unification occurs not only on a causal but also on a contemplative level:

> The One is all things and yet no one of them. It is the source
> of all things, not itself all things, but their transcendent
> Principle ... So that Being may exist the One is not Being, but
> the begetter of Being.

However, the One is not a primary source in a sense of 'the beginning' or 'the ultimate cause', but it is the 'eternally present possibility of all existence'. Plotinus asserted that the multiplicity of monads or psyches could not have emanated from a more complex entity; thus, it has emanated from the less complex entity of Forms (for instance, the idea of good and beauty in a Platonic sense), which ultimately emanated from the Simplex — the One. In other words, Plotinus implies that the origin must be simpler than what follows after. (The same view is central to modern evolutionary theory, and regarded as thoroughly scientific.) The exact same notion regarding the origin exists in the East, as expressed in the *Tao-Te Ching* of Lao Tzu (570–490 BC):

> Tao produced the One. The One produced the two. The two
> produced the three. And the three produced the ten thousand

things. The ten thousand things carry the yin and embrace
the yang, and through the blending of the material force *(chi)*
they achieve harmony.[38]

As to the characteristics of the One, Plotinus stated that: 'The One
as transcending intellect, transcends knowing ... Knowing is a unitary
thing, but defined; the first is One, but undefined'. (Plotinus, *Enneads*).
Again, the exact same idea can be found in Lao Tzu's *Tao-Te Ching*:

> We desire to understand the world by giving names to
> the things we see, but these things are only the effects of
> something subtle. When we see beyond the desire to use
> names, we can sense the nameless cause of these effects. The
> cause and the effects are aspects of the same, one thing. They
> are both mysterious and profound. At their most mysterious
> and profound point lies the 'Gate of the Great Truth'.[39]

Thus, the unitary understanding of ancient East and West reveals
that a higher understanding of the unity cannot be achieved only by
reason or rationality in a strict sense: rather, the unity manifests in
the whole of all possibilities of knowing, which our intellect finds
considerable difficulty in defining. Here again we might note that the
inability to know all is a limitation of consciousness.[40] (This limitation
is evidently not suffered by the omniscient 'gods' venerated by many
religions.) Put differently, the One is beyond intelligence in Plotinus'
terminology. Likewise, the Way *(Tao)* is the harmony of *Yin* (dark) and
Yang (light) — the harmony of the thousand things in Lao Tzu's terms.
The example of this rather 'arational' understanding of the unity is
observable in the Buddha who in the moment of *satori* (enlightenment)
said: 'I, the great earth, and all beings simultaneously achieve the Way',
expressing the boundlessness of the One.

These examples highlight some important aspects of eastern thought
and point to certain basic concepts of nature as understood by the early
Chinese. These concepts were articulated with such terms as *li* and
qi (chi) and many others. As explained for the western mind by the
Cambridge biologist Joseph Needham (1900–95) and colleagues in

their monumental encyclopedia *Science and Civilization in China,* these terms are very difficult to translate. According to the Neo-Confucianist Xu Zhu Xi (1130–1200), '*Li* is the Dao', the 'principle of organization', from which all things are produced. *Qi,* inadequately translated as 'air' or 'vapour', is akin to Hindu *prana* or Greek *pneuma* (or *psyche*) and perhaps bears some resemblance to what we now call 'matter-energy'. As stated by Zhu Xi: 'throughout the universe there is no *qi* without *li*, or *li* without *qi*'. They are 'cosmic principles of organization and matter-energy respectively' and, as such, they are, by their nature, expressions of cosmic unity and coherence rather than of separate things.[41]

In the west, Baruch Spinoza (1632–77) sought to 'understand' the reason for the unity in existence. In his *Ethics,* he rationally constructed a system that justifies the equation: nature is god.[42] Henceforth, he argued that : 'nothing can be or be conceived without God' (Proposition 15), and developed the proposition that the nature of existence is one, and the one is what Spinoza called God. For Spinoza, each attribute or a phenomenon (for instance, colour, extension, wonder and will) is an expression of God since everything is God. This idea is also immanent in Hindu tradition where gods and goddesses are perceived as aspects of reality that are united with the principle — *Brahma* (the Reality). As just noted, many modern scientists are also intuitively and professionally drawn towards exploring Grand Unified Theories (GUTs), and it is well known that Albert Einstein thought very highly of Spinoza's philosophy and concept of God.

However, the question for our rational mind is how the reality perceived by mystics can manifest in multiple forms if the so-called Reality is in fact the One. This question can be answered from different perspectives, both eastern and western. If we were to perceive the world like a Buddhist monk, we would perhaps see little or no boundary between phenomena since as they say: everything is connected to everything else. Thus, for Buddhist minds, there is less multiplicity and more unity in experience. This intuitive, perhaps mysterious way of perceiving is rather different from the western 'perspective' though this is also changing towards a more holistic paradigm. Buddhism tends to see differently, not just with eyes. In fact, the word 'mystery' originates from the Greek verb *muo,* which signifies to shut or close off lips or eyes.

Tantra, which emerged from both Buddhist and Hindu traditions, offers a somewhat more complex view using the Mandala or 'circle' (in Sanskrit) to represent an integrated pattern of spiritual reality either visualized or viewed. Mandalas represent the whole of an experience, a fourfold manifestation of the unity. Symbolically the Mandala integrates space and time, bringing together the four separate cardinal compass directions (in space) and the four seasons (in time) as well as many other inner/outer, conscious/unconscious and individual/ collective polarities (see Chapter 5).[43] Although tantra divides reality into four aspects, with such polarities as dual/non-dual, action/ meditation, it nevertheless acknowledges the wholeness of the reality, and the Mandala as a symbol of connectedness. The West, on the other hand, is traditionally much more analytical, perceiving the multiplicity of the One reduced to the level of atoms (Democritus), genes (Mendel), neurons (Waldeyer-Hartz), and so forth.[44] At the atomistic level, understanding the eastern (Buddhist) intuition of interconnected phenomena is more difficult. Science for instance discovered the neuron and studied it in isolation before understanding neural networks. The same can be said for genetics, which no longer stresses the supreme importance — or selfishness — of the individual gene, but instead recognizes their deep inter-relationships and the flexibility of genes within organic systems.[45]

These considerations are indeed perennial. The mathematician Henri Bortoft, a student of the renowned physicist David Bohm, has written a fine discourse on the subject of Unity and Diversity.[46] In addition to showing the influence of Bohm, Bortoft's perspective is an elaboration of the work of Goethe who understood that no scientist could be what he called a pure atomist or a pure dynamist, what today we would call a reductionist or a process- or systems-oriented holist. As the individual of any species grows from a single cell into an adult organism, it literally expands outwards (evolves) from its incipient simple undifferentiated state (unity) to a complex and more differentiated state, but as individual life draws to a close, a process of contraction (involution) sets in. The two processes are fundamentally connected, like the top of the wheel going forward while the bottom goes backwards. For example, as the spiral sea shell grows outwards, so its interior becomes more complex, and in morphological terms, shell biologists talk about evolute and involute forms (see Figure 2.4).

This process is equally obvious in the case of the plant (Figure 2.5) which begins with a simple seed in which all latent life forces are contracted into a very simple physical entity. Then as the plant grows and expands, a diversity of stems, shoots, flowers and fruits manifest before it again contracts into a seed. We shall show in the next sections that understanding this dynamic process is very helpful in understanding the dynamics of biological growth and changing consciousness structures, and that modern science is rapidly recognizing

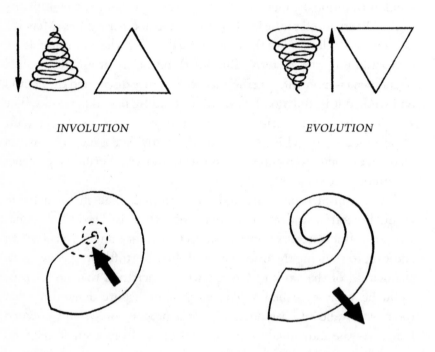

INVOLUTION   *EVOLUTION*

Figure 2.4. The reciprocal process of evolution and involution. Evolute and involute shells spiral outwards and inwards respectively. (Conceptual spiral motifs modified after Thompson 1981, p. 94.)

the importance of developing this dynamic new perspective. This process of expansion and contraction is highly organic and central to all life and to many other cosmic processes. Moreover, it is intimately connected to the concept of polarity between unity and diversity, especially as it pertains to the evolution of consciousness in the human lifetime: our consciousness emerges from unity, expands into a diversity of experience and contracts again into unity.

LEAF SEQUENCE IN PLANT

Figure 2.5. Plant growth shows an expansion-contraction cycle. (Modified from Bockemühl 1985, Fig. 2.)

Evolutionary theory postulates the expansive growth (evolution) of unicellular life into complex humans with evolved consciousness, but, typically, we humans question whether such simple ancestral cells could have any rudimentary consciousness. Likewise the Big Bang Theory postulates that all matter, all that is, originated from an even more unified and undifferentiated cosmic egg, eventually leading to human cosmology which becomes aware of the process, so that a physicist becomes the universe's way of contemplating itself.[47] The Hindu wisdom traditions intuited much the same cosmic history. Shiva inhales and exhales on a four billion year cycle, and three millennia later modern western science speaks of an expanding universe and an expansion of life on Earth which it considers to have been ongoing for about four billion years. Western science also predicts an eventual 'Big Crunch' (contraction) to match its 'Big Bang'. This just proves that with respect to the basic 'big picture' dynamic western science reaches the same conclusions as ancient Hindu cosmology.

In our exposition, our focus is to synthesize our understanding of western traditions of analytical, reductionist thinking and eastern traditions of synthetic, mythic-symbolic and holistic thinking into an integral view of consciousness. This will not be the first such effort. However, our attempt does not simply involve incorporating meditation and enhanced intuition with rational discourses, as many thinkers have advocated. Instead, we seek to show in concrete terms how biological processes are organic, not mechanistic, and so complementary to the development of our organic consciousness. We show how a new biology that integrates an understanding of consciousness at all levels, transcends the mind-body duality of the detached observer probing or explaining some other species. By transcending this self-other dualism, we gain a deeper understanding of relationship.

In its great outward expansion and 'exploration' of the external, visual, seen, world, which it has taken apart and reduced down to its subatomic constituents, modern western science has tended not to notice the shadow of unity that integrates all that is, whether we dismantle it or not. In noticing phenomena like globalization, the environment and the internet, we consciously become aware of the global and cosmic interconnectedness that wisdom and spiritual traditions have recognized for so long. But it is precisely these seeds of a more holistic and unitary worldview that herald the shift from the preeminence of reductionist thought towards a more holistic paradigm. The pendulum can only swing so far before the critical point is reached. Whether our metaphor is the phase shift to global brain status, or the inherent dynamic of a quantum leap from mental to integral consciousness, we, like all our global brethren, are part of the process. It is not merely our individual intellects proposing a new theory of evolution that other individual brains may wish to explore, question or even repudiate. This may be a small part of the process, but something much larger is going on. Consciousness is evolving, and evolution is becoming conscious of itself. In so doing evolution rewrites its own textbooks, in physics, biology and psychology. Newton, Goethe, Darwin and Laszlo are merely the scribes, the organs that develop their potential and fruits as intrinsic parts of the process.

This idea of seeing the whole rather than the parts has also been gaining increasing attention in the West, and not surprisingly because many are anxious and even afraid of the alternatives: social and environmental fragmentation and discord.[48] In their little book *Anxiety,* Heiri Steiner and Jean Gebser noted that such anxiousness precedes a phase shift in consciousness.[49] It is the perennial issue of growing pains, which may, in the short term, create unwanted discord and confusion before the phoenix arises from the ashes. The death of the old system may even create a 'Dark Age' before the new age is born.[50] So it may be that we are simply doing what it is time to do: dismantle the obsolete machinery before we reunite those parts of the diverse whole that have become disconnected. Fundamentally, our synthesis is a gravitation towards bringing together the eastern and western or mythical and rational consciousness traditions and so creating a new integral synergy. In this process our self-consciousness shifts and with it the centre and the boundary. Our inner and outer worlds change, individually and collectively. The river of consciousness flows on and the adventure continues.

PART 2.

The Evolution of

Consciousness

3. Consciousness and the Individual Life Cycle

Being childish: how our minds grow

'Don't be so childish!' the exasperated parent exclaims as he/she tries to constrain a child's behaviour to conform to the conventions of an adult situation. Occasionally, the quick-witted child might reply: 'But I *am* a child.' The sensitive parent will hopefully appreciate the child's wisdom and realize their error. Here, in a nutshell, is a clash of cognitive expectations — an encounter between two different consciousness structures.

So, consciousness is not static. It changes, as we grow. Even as adults, when we may become somewhat set in our ways, and comfortable with certain mindsets, we still often aspire to broaden our horizons, to expand our consciousness. In childhood, our consciousness changes very rapidly as our infant horizons expand. Moreover, this happens as part of an intrinsic biological process, without any self-conscious desire or drive on the part of the individual. Growth is a powerful, unconscious and wilful process. As any child development psychologist will tell us, such growth is both physical (biological) and psychological or cognitive. Among those we most have to thank for pioneering this field of study are Jean Piaget and Sigmund Freud, both founders of the modern science of Childhood Development.[1]

Like Piaget, who we shall examine shortly, Freud (1859–1939) stressed that cognitive development is a dynamic process.[2] Freud divided the human mind into three domains, structures or apparatuses: the Id (biological needs), the Ego (conscious), and the Superego (conscience or higher self), each with its own unique functions and priorities.[3] As they develop dynamically in any healthy individual, the more primitive (unconscious) psyche gets obscured (repressed) as newer, more conscious states emerge. Freud's observations may seem familiar, obvious and intuitively correct with a hundred years of psychological hindsight under our belts, but it is worth reflecting on the importance of his insights. First, he noted that as one consciousness structure dethrones another, the previous one is not lost, only repressed. Second, Freud saw his threefold schema as fundamental to the operation of the mind. Thus, the Id, the most primitive (unconscious) psychic structure, which is driven by self-preservation desires and needs, is always in conflict with the faculty of Superego, which operates from the higher plane of self-criticism and morality. This leaves the Ego in the 'middle' in a dynamic tension between the two poles, trying to accommodate potentially conflicting or antagonistic needs arising 'above' and 'below'.

According to Freud's rather pessimistic view, one state of consciousness replaces another in: 'a tale of disappointment where each disappointment leads to an alteration in the I'.[4] Awareness of such changing psychological dynamics is the first step to inquiring deeply into human existence and its meaning: 'to reclaim a part of nature for oneself, allowing the boundaries of the soul [to be] redrawn so that what were once taken to be forces of nature I now recognize as my own active mind'.[5] Thus, humans can master themselves by taking responsibility for their own being, through *individuation*. This may be more easily done by the mature-minded adult than the impressionable child, but Freud's stated principle is fundamentally important. We change, in a dynamic process of becoming: so '[w]here Id was, there shall Ego be'.[6] According to Freud, the psychological drama of existence involves the transformation of mere impulses into conscious decisions. Put another way we grow to learn and accept our biologically-determined nature and then set about freeing ourselves by gaining mastery over it. But what is scripting the drama in the first place?

What is important for us to understand here is that Freud pioneered a model of development (ontogeny) that accommodated 'different' states or structures of consciousness that coexist in the individual —

with different emphasis at different times. So if we say to an adult, 'Don't be childish!' what we really mean is adopt a rational cognitive posture suitable to your adult age (and don't revert to an early infantile consciousness structure). Freud further suggested that conflict between the two poles of Superego and Id is what led to imbalance and pathology in his patients. But, as we shall see, psychological systems are dynamic and so may sometimes vacillate between balance and imbalance. In addition, like Carl Jung, Freud believed that the Ego manipulates the language of symbol and active imagination to express/manage the psychological discomfort and conflict that arises.[7] As we shall see, Freud deserves much credit for describing the threefold structure of the mind, and we can build on this model in very fruitful and constructive ways. However, before doing this we should turn to the classic work of Piaget.

Jean Piaget (1896–1980) is particularly famous for his studies of childhood cognitive development. A scholarly, renaissance genius with wide-ranging interests, at various times he simultaneously occupied university chairs and important institutional positions in psychology, sociology, history of science, education and epistemology. Most summaries of Piaget's work outline four important stages in childhood and adolescent development.[8] His now-famous and oft-quoted scheme makes cogent sense to anyone who has taken the time to observe that what really goes on in childhood development is a fascinating example of the evolution of consciousness. As a result, Piaget's wisdom has now become incorporated into popular books on child rearing and should be required reading for any thoughtful parent, who realizes the potential dangers of not understanding that children behave like children. We will briefly discuss Piaget's stages of cognitive development:

1. Sensorimotor stage
Between birth and about the age of two, the very young infant develops through the sensorimotor stage. At this age the child is busy developing its relationship with the physical world through interaction with the environment. Activity mostly involves sucking, grasping, gripping and making noises. Specialists use terms like 'prehension' for grasping and 'vocalization' or 'phonation' for making noises. This is also the

'peek-a-boo' stage, when the child may show genuine delight with the appearance and reappearance of the peek-a-boo player's happy face. This is not just because the game is fun, as might be the case with an older child, but because of what is called 'object permanence'. Objects are perceived as temporary and transient, so the child genuinely does not 'know' that the hidden person has not completely disappeared. Thus, the reappearance of the familiar face is a stimulating experience. Conversely, the child can be genuinely distressed when a parent disappears because they genuinely believe them to have disappeared completely. In psychological jargon we call this 'separation anxiety', and for infants it is very real, immediate and upsetting. It is obviously no coincidence that peek-a-boo is a game played with the face — the seat of the visual and vocal sense organs and the conduit for our most expressive, communicative gestures (smiles, raised eyebrows, winks, and so on).

Anatomically, the head is the most developed part of the body at this early stage of development, while the body remains very uncoordinated. An excellent demonstration of this fact is provided by the German specialist Karl König who outlined, in great detail, how the child develops the ability to lift its head after about three months, to sit up at six months and stand up after about 9–12 months. Thus, the child grows into its body from head to toe. This development can be described as the flow of physical or motor consciousness into the body beginning with the head.[9] (See Figure 3.1.) In technical terminology, this head-to-toe development is referred to as an anterior-posterior (or cephalo-caudal) growth dynamic.[10] Simply put, we come into the physical world head first, not only literally, as in most normal births, but equally literally with the head most-developed. Then, the flow of motor coordination from head to toe allows the head, trunk and finally the whole body to come into a new equilibrium with Earth's gravity. (As we shall see, the anterior-posterior current of growth does not complete development of the physical body until well into adolescence.)

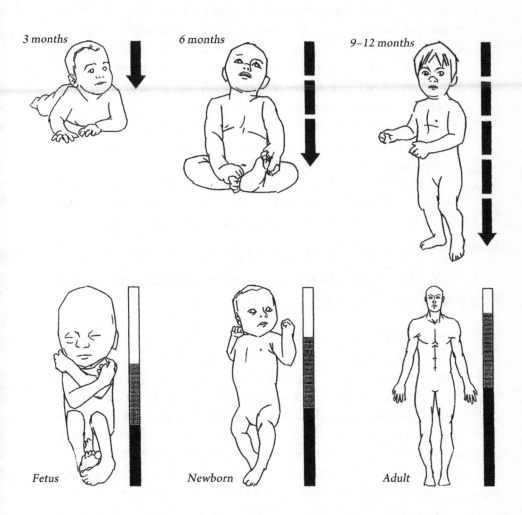

Figure 3.1. Head-to-toe growth trends realize a huge relative enlargement of the trunk and limbs in proportion to the head (lower sequence). The upper sequence shows how motor control begins with the head, then trunk, then limbs during the first year of life.

If anyone has ever wondered why young infants grab objects and throw them around, one only has to try and identify with the extraordinary sense of power that the child experiences as it realizes (quite instinctively) that it is establishing a relationship with the environment and gaining control of physical objects. Thus, it will bang on the table and throw things down from a high chair, and do so *repeatedly*. Having just come to terms with their own physical body, it is only natural for children to explore the objects within reach and establish relationships with them. By throwing things on the floor *repeatedly,* one explores the immediate space around, and establishes that the floor is *repeatedly* found where thrown objects hit it. In adult terms, the child is experimenting, even conducting a *repeatable experiment*, but not in a conscious, cognitive sense. Repetition is very important for children, and something they do instinctively. Clearly, even as adults, we cherish the security of familiar routines. The fractal organization of the universe involves a rhythmic alternation of seasons, orbits and life cycles. The 'three Rs' of repetition, rhythm and recursion are part of the deep structure of existence.[11]

2. Pre-operational stage

The next phase of development, which Piaget calls the Pre-operational Stage, between about the ages of two and seven years, represents an inward shift from physical groping to mental groping. In its ongoing physical groping, the child cannot yet conceptualize abstractly and needs concrete physical situations to help it relate to the world. Tactile activities are important for healthy development. Kids at this stage love to touch things, play with clay and food, and find out if they are hard, soft or 'squidgy'.[12] For this reason adults frequently intervene with 'don't touch, don't play with your food,' although from a developmental standpoint this is often the wrong message, unless the child is in danger of doing itself harm. Parents and kindergarten teachers recognize the child's need for physical contact with the world and so provide it with many hugs, and constant opportunities to play with toys and other objects.

Obviously there are vast differences between the young two year old and the seven year old, and we could easily look at many subtle and fascinating stages of development during this formative five-year period. Many of these developments can be outlined in terms of the mental groping metaphor, which we observe as the child uses language to ask endless questions or describe what it sees in its own inimitable way.

We have seen that repetitive actions, like banging on the table, help establish a physical relationship with the world. As noted, imitation of adults, or other children, is important, and is in itself a type of repetitive behaviour. Indeed the stereotype of play is that children imitate the adult roles of parents, doctors, nurses, teachers, actors, magicians, and so on. They may equally well adopt the roles of dinosaurs and dragons. But the play-acting is real for the child in a way that it is not for adults. The child believes it is a scary dinosaur, and believes it is hidden in the bushes, when Mom can see it perfectly well behind the kitchen counter. Just as the young three or four year old may be genuinely frightened by a large dinosaur sculpture in a museum, especially the moving robotic type, so too the child may believe that Mom is really scared when ambushed in the kitchen by *T. rex*. So the child 'participates' fully in acting out roles, and believing in both real and imaginary figures. Like cartoon and fairy-tale characters, Santa Claus is also real.

The ability to believe in such fictional characters, or to talk to dolls and teddy bears as real animated characters, can be labelled as Animism. Indeed, the child easily believes that inanimate objects are alive. The pencil is alive because it writes, and fire is alive because it grows. The sun is alive because it watches us. The child may believe the sun is alive because it travels across the sky, or follows the child home at the same time as the pet dog. Things have their own magic animating spirits. A glass knocked off the kitchen counter 'fell off by itself'. Similar 'thinking' was characteristic of early Greek culture. Physical objects were thought to have souls *(anima)* because they could move.

At this 'Pre-operational Stage' the child does not understand abstract principles such as 'equivalence'. This affects the sense of space or spatial distribution. So, in one of Piaget's classic examples, a given amount of water in a low, wide glass is seen as less than exactly the same amount in a high, narrow glass. Even when the child sees the water poured from one glass to the other, the apparent increase is perceived as a real (magic) transformation from less to more. The magical consciousness structure is important at this stage of development as the child very fully participates in the magic of transformation of self and playmates into dinosaurs, super-heroes and the like.

Another fascinating example of spatial perception can be seen in drawings. Pre-operational children either just scribble when aged between about 2–4 years, or, more typically, between the ages of about 5–7, they draw human figures with large heads and disproportionately small and sketchy bodies (see Figure 3.2).[13] This reminds us of the inevitable focus on the head as the seat of the sense organs, but also has interesting resonances with the very 'real' biological fact that one does not fully grow into one's adult body until around the age of 21 when all phases of tooth development and final ossification of limb bones are complete.

Incipient space and time consciousness also develops in such activities as counting and writing out the letters of the alphabet. However, these activities maintain their strong repetitive characteristics, so typical of this stage. Moreover, they are animated by being repeated aloud, with many examples: A for apple, B for bumble bee, C for cat and so on. Only in this way are abstractions like letters given meaning. Repetition is a staple of primary school learning as the whole class repeats the names of dinosaurs, mantra-like multiplication tables and other key learning phrases. Such exercises force abstract representations into consciousness. Even as adults, we find repetition is important, especially in activities like driving and sports, as a means of ingraining activities into the psyche until they become 'second nature'.

Just as the space occupied by the child is flexible and easily transformed from kitchen to Jurassic Park, or from closet to cave, so too the child's concept of time is flexible and quite different from that of the adult. One of us once witnessed a classic exchange between mother and son (aged 3 years and 11 months). As the boy was leaving to go to pre-school with his Star Wars light sabre, mother said: 'You can't bring that today — show it and talk about it tomorrow.' The reply was quite simple: 'But today is tomorrow.' At this stage the child may say that Dad is older than Grandad because Dad is more important. By the

Figure 3.2. Drawings of humans by a young child (left) and an adolescent (right). (After Goodenough 1926, Figs. 9 and 20.)

same reasoning Dad may be older than God (and may even feel that way on occasion.)

A sense of morality (the Superego) also begins to emerge at this stage. Things may be considered good or bad, based on outcome and the reaction of others. Recognition of rules necessary for games will also develop, although the rules may not be obeyed with any consistency. The child will probably be unable to distinguish between a bad outcome that arises as the result of an accident, and one caused by a deliberate action. So breaking a glass is bad even if it is an accident, and Mom consoles by saying: 'It isn't your fault.'

3. Concrete Operations stage

Around the age of seven, many important shifts in biological and psychological development occur. As is the case with all these shifts they do not occur at exactly the same time in all individuals, nor do they represent complete or absolute changes in consciousness structure. The shift from Pre-operational to Concrete Operational involves the emergence of thought. Since adults think, we are inclined to assume young (pre-7 year olds) also think, and maybe they do to some degree. However, there are strong arguments against the position that it is the same type of thinking that we adults refer to. For example, we do not hold young, pre-operational children responsible for decisions involving moral and ethical judgment. Indeed we do not hold them legally responsible until they are much older, as discussed in reference to later developmental stages.

Between the ages of seven and eleven, we see the development of the so-called Concrete Operations stage. As physical experience accumulates, the child starts to conceptualize, creating logical or concrete 'thought' structures with which the child explains his or her physical experiences. Abstract but simple maths problem-solving becomes possible at this stage without having to count actual objects. But conceptualization may remain limited. The seven-year-old child says he has a brother but if asked if his brother has a brother he may say no. Can this child really 'think'? Morality at this stage begins to become more cooperative. Children make up rules to maintain imaginative flexibility, but still agree to adhere to these rules to avoid quarrels. But, according to the pundits, there is still a lack of 'moral realism'. Thus, breaking two cups is worse than one regardless of intent.

The sense of space and time is far more developed in the 7–11 year old than in earlier developmental stages. Artwork at this stage can be quite developed, and in typical examples of figure drawing we have seen realistic depictions of the proportions of head and body, often with fine detail of clothing and other attributes (see previous page, Figure 3.2).[14] If nurtured, these artistic abilities probably improve in subsequent stages of development. We may also take it for granted that the ability to read and write becomes increasingly well developed at this time. This stage of development involves combining improved artistic ability with a more mature sense of space and time. The young child takes time to exhibit 'well developed' handwriting, and it is interesting that we refer to scribbling for undifferentiated drawing and writing. The ability to link different symbols (letters) together and give them meaning, as words, is a good manifestation of concrete, logical thought. The letters C A T are an entirely abstract representation of the family pet, and until the representation is understood by consciousness it has no meaning.

A child's understanding of time is also much more highly developed at this stage. To continue with the example of writing, one must have a sense of time in order to link words and their meanings into a story: the cat sat on the mat. Although the pre-operational youngster may well get this far before the age of seven, as we shall see in the next section, some educational philosophies suggest postponing teaching of writing and reading until the age of about seven so that other developmental stages are not short changed. Regardless of the subtle distinctions between writing abilities of children prior to, or after the age of seven, they are all likely to develop a worldly sense of time at this stage. Indeed, in most cultures they will be off to school where they are expected to know when classes start and how to read the symbolic hands of the clock face, or the numbers of the digital display.

4. Formal Operations stage

Emerging around ages 11–15, the child's cognitive structures begin to take on adult characteristics that include conceptual or hypothetical reasoning. Thus they are capable of ruminating on 'what might happen if no one never died?' Or 'how would someone else feel if I was rude to them?' This helps explain why young adolescents may be unusually sensitive to the feelings of their friends, and the inconsistencies, apparent or real, that they perceive in the rules laid down by parents and authority figures. The parent may put this down to 'individuation', and it is indeed true that all Piaget's stages of development involve various, progressive degrees of individuation. Researchers like Eric Erickson refer to the identity versus confusion stage, and infer that establishing 'self-identity' is a necessary pre-condition for the development of abstract thinking.[15] At this stage, therefore, the consequences of conceptual and hypothetical reasoning have their own very specific consequences for the adolescent's emotional and cognitive development and self-image. So, as moral realism develops it is possible to better understand intent, and recognize the inherent morality in society's laws and adopt an adult approach to responsibility. So, at least in countries like the USA, it is legal to leave a twelve year old 'home alone'.

On a more strictly intellectual-quantitative level the adolescent is becoming a scholar capable of exploring proportions and ratios. How many possible combinations are there of five things? At this stage it is possible to move beyond simple arithmetic to do algebra involving hypotheses: if a = b, and if x = y.

The rhythm of life: intellectual and emotional cycles

Piaget's classic model of child cognitive development is revered as a model of insight. However, by definition, it only deals with early ontogeny. As the child matures 'progressively' into adulthood it is easy to conceive of this ontogeny as a linear process in time. But, is such a linear view of time satisfactory? Our direct experience of time is not only linear, as one day follows the next, but also cyclic as days biorhythmically follow nights, and build into the cycles of lunar months and the seasonal cycles of the solar year. We know that the female reproductive cycle is strongly influenced by lunar cyclicity as are the reproductive cycles of many other species.

In this section we shall examine the seven year cycle. As we saw in the previous section, Piaget identified seven as a pivotal age in cognitive development (between the pre-operational and concrete operational stages). However, it is also a biologically significant age when the replacement of baby teeth is completed by the eruption of the first molars. The second molars appear and move into position between the ages of twelve and fourteen, and the third molars — the wisdom teeth — appear around the age of 21, when the final ossification of limb bones is complete.[16]

These three seven year stages of physical development are accompanied, and later followed, throughout life by more subtle psychological developments. Several authors have noted this, and some even claim to have been able to identify such cycles of development in the self-portraits Rembrandt painted throughout his life.[17] The seven year cycle was considered important by Rudolf Steiner as a natural guide to emotional and intellectual development.[18] Like Piaget, he considered that prior to the age of seven, the adult mode of thinking was not really well developed. He applied this principle in his Waldorf schools, which avoid forcing children to develop cognitive skills prematurely. Thus, he advocated that seven was the appropriate age to learn to read, when early stages of emotional development had been completed naturally and healthily. It is highly pertinent to our previous discussion of the child's early development to note that Rudolf Steiner emphasized the fundamental importance of physical activity (play) and the constant use of the limbs for healthy development. It is only too apparent that lack of physical activity in childhood often causes developmental problems. Also, as Steiner realized, it is damaging to children to subject them to intellectual trials and tests before they are ready. He would no doubt have been appalled that we have the temerity to give children powerful drugs like Ritalin to stop them fidgeting at the very time in their development when they should be exploring the world with perfectly natural hyperactive enthusiasm.

A good overview of the seven year cycle was presented by William Irwin Thompson in his 1973 book *Passages about Earth*.[19] There he characterizes the cycle as a seven year alternation of emotional (*Homo ludens*) and intellectual (*Homo faber*) development that proceeds until approximately the age of 35 years: one might say until early mid-life. After this the cycle attenuates to fourteen years (35–49 and 49–63), at which point the human attains the mature status of *Homo sapiens*. This suggestion of the late development of wisdom may seem rather vague, symbolic or whimsical. However, it is resonant with the more general notion of the maturation of the developed human being towards the stage of 'sage' or 'cosmic consciousness' later in life. Human life is rich in emotional and intellectual experience. As the more physically dynamic stages of development prior to mid-life give way to the post-reproductive (post-menopausal) stages, so mature experience takes on new dimensions as parents become grandparents and even great-grandparents, often at about the time they lose their own parents and become the elder generation. In many cases, the afore-mentioned '*sapiens*' wisdom is clearly manifest in life-changing shifts in conscience and consciousness which come from mature engagement with the world community.

In investigating these cycles, we found a very similar scheme recorded in the Ashrama system of the Hindu tradition.[20] Here life is divided into four stages as outlined in Table 3.1.

In this scheme the later stages of life are associated with a spiritual and moral maturation. As we shall see there is evidence from western studies that spiritual or religious epiphanies, which often happen quite spontaneously, are often a characteristic stage of development in mid-life. Those who have studied and experienced such epiphanies often regard them as spontaneous 'evolutionary' shifts towards cosmic consciousness. Before undertaking an analysis of the evidence for such intriguing shifts towards what might be called 'higher' spiritual consciousness, we should first explore the earlier stages of the *ludens-faber* cycle that we all experience as we grow up.

In the western tradition, the *ludens-faber* notion can be traced back to the work of the French philosopher Henri Bergson, whose influential book *Creative Evolution* (1907) helped him win the Nobel Prize in literature. His notion of *Homo faber* — the builder or fabricator — comes from the well established notion of man the toolmaker, repeated in countless archaeological texts, and also used by Karl Marx. This popular notion caught on in many circles and there are at least two books entitled *Homo faber*: the first, written by G.N.M Tyrrell in 1951, deals with human mental evolution, while the second, written by Max Frisch in 1957, is a novel.[21]

However, as the old adage goes: 'all work and no play make Jack a dull boy'. We humans do not do well if we have to work constantly either on a physical or mental level. We like to relax and play, or enjoy what we sometimes refer to as 'mindless' pastimes when we 'veg out'. (Presumably vegetables don't think too much!) So, the compensatory side of the constructive *faber*- or work-mode is the *ludens*- or play-mode. *Ludens,* from which we get words like ludicrous, speaks to the emotional side of life. As modern psychology informs us, a balanced personality is able to integrate both emotional as well as intellectual intelligence. The longer we work or focus on a constructive *faber* task the more we look forward to a break and the emotional release of relaxation — a change of pace. On the other hand if we relax for too long, we sooner or later want to do something constructive. Indeed, in ideal circumstances our work day, work week and even our working year is divided into just

Table 3.1. The four stages of life in the Hindu Ashrama system.

Life stage	Age
Celibate student years	7–21
Artha Stage or householder years	21–40
Dharma stage: issues of meaning and truth	40–70
Moksha stage: liberation spiritual fulfilment	70 +

such cycles. However, the notion of the *ludens-faber* cycle speaks to intrinsic rhythms of the life cycle that are far more fundamental and biologically ingrained than any we impose by organizing happy hours, weekends and holidays. Nevertheless, although intrinsic biological and physiological cycles have certain universal characteristics, the following examples of typical changes in behaviour from cycle to cycle are taken from the experience of western society, and it should be evident that the experience of individuals in certain other cultures would be different to varying degrees.

The first ludens phase (up to age 7)

As noted by Piaget and endorsed by Steiner, for normal healthy development the first seven years of childhood should be a time of playful exploration in the world. This is the first *ludens* phase: the pre-school years before the mind is oriented towards serious thinking. Reading, writing and other intellectual pursuits are treated as games that are easily abandoned if they are not fun. However, at this stage when verbal communication is so important, children rapidly pick up languages, without ever seeing or writing a word. Young children and their parents talk of 'playing games' rather than 'sports'. The child is

Table 3.2. Human emotional and intellectual development charted on a modified seven year cycle (after Thompson, 1973).

Developmental stage	Age in years	Developmental stage	Age in years
Homo ludens	1–7	Homo ludens	28–35
Homo faber	7–14	Homo faber	35–49
Homo ludens	14–21	Homo ludens	49–63
Homo faber	21–28	Homo sapiens	63 onwards

still developing his or her ability to socialize with other children and family. The child can be happy and emotionally secure with parent or friend one minute and distressed the next, because of accidental changes in circumstances that cannot be rationalized. For example, the child is easily frightened or emotionally elated by the moment to moment unfolding of daily events: the appearance or disappearance of family, friends, pets, and so on. Children often respond with great empathy to the emotional states of others: 'Don't be sad, Mommy.' This is a great time for parents to involve children in 'feeling' for others, taking them presents on birthdays, or when they are sick, and helping to cheer up Grandma by paying her a visit. Likes and dislikes are explored and based on emotional impulses not on any intellectual or rational process. Life revolves around home base, which provides the child a secure grounding in the world. From this emotionally secure position the normal child will often ask its first essentially religious, spiritual or metaphysical questions such as: 'Where did I come from? What happens when I die? What does God look like?' Mostly the child has no need of a rational answer, as long as the question is met with an empathetic response.

The first faber phase (age 7–14)

This is the elementary or primary school stage prior to adolescence. The normal child begins to take a serious interest in facts that pertain to the external world in which it now becomes an active participant. Traditionally this has been the model-building phase in a boy's life. As he constructs a model of an airplane or racing car his head is full of facts about its dimensions, speed and other attributes. He may know more than his adult peers about the height of Everest, the distance to the moon, the length of the Golden Gate Bridge, or the weight and hunting behaviour of *Tyrannosaurus rex*. He may also become an expert on sports statistics and schedules, and indeed may be a stickler for getting relatively trivial facts right, and for demonstrating his new-found mastery of knowledge to others. Girls may be equal sticklers for factual truth as it affects their lives and interests. Although girls may traditionally be more emotionally mature and more preoccupied with the domestic scene, their inclinations are probably still focussed on organization, whether it be of chore schedules, food preparation, clothing or relationships with siblings and school friends. Whereas a four or five year old will indiscriminately 'want' and grab food items in a supermarket, unable to understand that the store is not a huge playground or toy shop, the eight or nine year old can help select items from a shopping list, rationalize the preference of one selection over another and probably organize the dinner menu. In the present age of computers and cell phones, boys and girls at this age become adept at operating these communication devices which not only give them access to the external world, but do so through the medium of organized menus and data displays.

The second ludens phase (age 14–21)

Just as the replacement of baby teeth by around age seven represents an important stage in biological development, so the eruption of the second molars, which coincides with the process of sexual maturation, also represents a permanent landmark in the child's relationship with the physical world. Puberty may begin before age fourteen, and indeed

The God Joke
A child is drawing a picture in elementary school and
the teacher asks:
'What are you drawing?'
'A picture of God,' the child replies.
After a moment's thought, the teacher replies:
'How nice, but no one knows what God looks like.'
'They will when I've finished my picture!' comes the
emphatic response.

in recent history its onset is occurring earlier and earlier, but it is
nevertheless intimately tied to the onset of the second *ludens* cycle.[22]
In western society, the sometimes rather unfortunate stereotype of
adolescence is one of uncontrolled hormonal flux, pimples and rebellion.
As children become biologically-mature, sexually-reproducing adults,
they may expose themselves and others to sex, drugs and rock and
roll, all of which are likely to be more easily found as the student
graduates from elementary to secondary school, and later to college.
In many traditional archaic cultures, this transition was or is marked
by initiation rites, allowing men, to participate for the first time in
hunting and other more or less sacred rituals involving tests of physical
endurance, or sometimes in sexual rites or consciousness-shifting
psychotropic drug ceremonies. In western society we still maintain
many such initiation ceremonies, if only in somewhat diluted form.
It is interesting that certain 'spiritual' or religious initiations such as
confirmations and bar mitzvahs more or less coincide with the start
of this cycle, whereas others of a more secular nature, such as the legal
right to drink alcohol, drive, stay out late, own property, vote or join
the armed services, are instituted later towards the end of the cycle. All
of these potentially involve shifts in consciousness and the individual's
relationship to family and society. However, these cultural traditions
have evolved over the years, and in most modern societies they are often

not consciously thought of as initiations that coincide with the very real biological and psychological changes that accompany the emotional and intellectual consciousness shifts of the *ludens-faber* cycle. Likewise, while many mothers and daughters certainly treat first menstruation as a biological and psychological landmark, they are unlikely to use the *ludens-faber* language of ontogeny, or place puberty in the context of the next five seven year cycles that ultimately link puberty with menopause.

With sexual maturity comes a new physical prowess, which is often manifest in athletic excellence, and the ability of the child to out-do the parent on the sports field. (Any athletic parent will probably remember the day that a son or daughter outdid them in some physical activity.) We may refer to such athletic activity as a 'healthy outlet' for 'raw' or untamed adolescent energy. Instead of rebelling against parents or society, the young athlete can 'break' records, and help to establish new norms or standards, much as he/she may do with music, art and in other creative and emotionally charged arenas. Such rebellion although it may be healthy and creative, is not always seen as such, especially if there is extreme physical risk as in car racing, and teachers and parents may try to push adolescents towards safer, more intellectual, academic pursuits. When this happens there is a danger of intensifying the rebellion and encouraging adolescents to take sides with their peers with whom they identify on the emotional level. The stories of most pop culture icons from the Beatles to the latest heroes of the rap, skate and snow board circuits begin in early adolescence with creative rebellion against the existing norms. Thompson has pointed out that, especially in America, high schools and junior colleges are glorified sports factories. There is good reason for this. Many athletes reach their physical peaks between the ages of 14 and 21 and in many sports the chance of winning Olympic gold is greater in this seven year age bracket than in any other.

The second faber phase (age 21–28)

Developmental psychology is a complex business, but to keep our commentary relatively simple we summarize the first three seven-year cycles as more or less coincident with the phases of pre-school, elementary and secondary education, culminating respectively in the eruption of the first second and third molars. So, there is a broad correspondence between our physiological cycles and progression through typical educational systems in the secular world. This parallelism between education and development may continue as the opportunity for 'higher' university education becomes available in the second *faber* stage. A few historical records and even some contemporary reports document rare cases of precocious geniuses who attended university anywhere from the age of about ten to fifteen, which we have bracketed as belonging to the first *faber* cycle, or the early years of the second *ludens* cycle. But, it is of great relevance to the present discussion to note that most such stories of precocious intellect are accompanied by cautionary tales about the emotional vulnerability of the individuals in question. In all such cases it helps to remember that it is possible to impose development that is out-of-sync with natural development (ontogeny), and that the developmental trajectory is by no means the same for every individual, and can be modified by any number of circumstances. Nevertheless, as Piaget has so ably demonstrated, there are general rules pertaining to development, and if they are thwarted, compensations take place leading to emotional-intellectual imbalance and the need for adjustment. For example, the adolescent forced to study against his/her will in high school may become a wild, 'delayed adolescent' party animal when let loose in college, whereas the graduate who takes a year or two off to travel and see the world, may willingly return to university as a mature young adult keen to master the new skills required for completing various stages of higher education.

If we focus more on the general rule rather than the rare exception, we typically find that before the age of 21 the individual, despite being sexually mature by age 14 or younger, has not become fully adult in all aspects of physical growth, and is usually not regarded as an adult in the eyes of secular society and its laws. Until about 21, although age varies considerably, young adults may still live at home, at least part time, and are unlikely to have committed themselves in any type of profession or life career. Also despite considerable variation between cultures and individual cases, it is rather unlikely that many individuals are married with children at the beginning of this stage, although the reverse may well be true towards the end of this stage.

If we take the university student, young parent or young professional as representative of this stage of development, we see how they have the opportunity to channel their recently adolescent idealism and creative energies into responsible engagement with the institutions of society in the so-called 'real world'. Whether they be law students, Peace Corps workers or young parents establishing a home, development at this stage of life is fast becoming a fully adult proposition, largely independent of the authority of home and school that held sway in the years of childhood and adolescence.

The third ludens phase (age 28–35)

At this stage of life the young adults have gone through the first phase of fully adult development, what Jungians call self-formation. They may well have established themselves as home-owning residents of new neighbourhoods where they can develop social ties based on their professional lives and the demographics and dynamics of the communities in which they live. They may well have a new status in society, having qualified as doctors, lawyers, teachers, managers of various departments or even as successful politicians, CEOs, or celebrities, perhaps with access to more or less disposable income. The period of professional apprenticeship may be over, and may give the individual a confident sense that they have arrived, and can take on the world. The sense of struggling to prove oneself may be diminished, allowing a playful relationship with young colleagues and friends in the social milieu. As young parents, individuals will probably be refurbishing homes, raising young families and building new communities as they introduce their children to one another. Often, more importantly, the parent will enter the emotionally charged word of the child and enjoy the fun of baby talk and childish play. At this stage, physical and recreational activity remains important. Individuals are still young enough to enjoy physical strength and athletic coordination and to participate in sports leagues and community events, for fun, rather than with unrealistic Olympic aspirations. They can also join in their children's eagerness to explore a world of parks, museums and vacation destinations that they themselves had perhaps had less time to enjoy during their hard-working professional apprenticeships.

The third faber phase (age 35–49)

By traditional reckoning, age 35 is the midpoint in the tally of three score years and ten. Thus, 35 is a convenient marker for the shift from self-formation to self-realization and, as we shall soon see, it may mark the season of epiphanies. Even the optimist would not place mid-life after 49, which happens to be the average age at which women reach menopause, and also about the age when there is a universal deterioration in eyesight for both sexes.[23] Not only is one getting older, but children and parents are getting older also. So in mid-life one is sandwiched between the generation that created you and the one you created, with the potential for significant family and financial responsibilities and tensions. All things being equal, professional and domestic responsibilities are as significant as ever, and 'work' is part of the daily vocabulary.

It is perhaps no coincidence that, according to Thompson's model, the seven year cycle has now been attenuated to a fourteen year cycle; time seems to pass more rapidly, if not for the individual, at least for the rapidly growing children, and the aging parents. As a responsible party, viewing the family life-cycle from a mid-life 'sandwich' perspective, one may, perhaps for the first time, gauge the whole timeline of existence with new perspective. It is also no coincidence that individuals may undergo a 'mid-life crisis' at some point in this developmental stage, though it may happen at other times, or even more than once. This crisis may be of a spiritual nature and may be more or less disruptive of the life of the individual, and his/her friends or family. But, this does not necessarily mean that it is a negative experience, and it may lead to what is sometimes loosely referred to as 'reinventing' oneself. Even when accompanied by the disruption of break-down, divorce or illness, the result may be a rebirth and renewed emotional, mental and spiritual health for the individual. By such a yardstick, a mid-life crisis is not a mere pop psychology label to apply to deviations from conventional behaviour. Such crises may involve significant shifts in consciousness. As we shall see below, many of those who experience such crises may describe them as epiphanies or revelations that change careers and even the course of their lives. Although the causes cannot easily be categorized, they may be said to be necessary, natural developments, that cause a shake-up in consciousness which may arise when individuals have pressing conscious and unconscious needs for such radical change. This may occur in situations where it has been impossible to bring about necessary changes more gradually.

The fourth ludens phase (age 49–63)

Somewhere during this stage of development, if it has not happened already, it is more than likely that one or more of the following events will take place. One's parents will die, one's children will leave home and one will become a grandparent. One is no longer in the middle of the family sandwich, nor does one have the same responsibilities for the generations that preceded and those that follow. More than likely it is time to think of retirement and perhaps downsize to some degree, passing family archives and heirlooms 'down' the family tree. It may be time to take up a hobby, craft or pastime that has been neglected, or to re-enter childhood by playing with the grandchildren. It is certainly a time when ambitions to conquer the world and exercise power are likely to give way to concerns about health, private life and a more compassionate engagement with the community. As Thompson points out, just when the first serious thoughts of old age and death start to register, one may be rescued by the new life and energy of grandchildren and a 'fuller and deeper humanity' born of mature experience.

The final sapiens phase (after age 63)

This, obviously, is the age of retirement, recognized by society's pension schemes and the Beatles' famous song: *When I'm sixty four*. According to Thompson, it is the time when the human deserves the label *sapiens*. No longer a mere builder or player, the individual can hope to have developed wisdom. In our scheme of simple, self and sage consciousness, the sage stage has been realized: the *moksha* or 'liberation' stage in the Hindu tradition. This is just a representation of desirable maturity, and there can be no doubt that many would argue for manifestations of wisdom long before age 63 and, conversely, manifestations of foolish ignorance long after this age. But in the most general terms, having given the best part of one's life to family, profession and community, one has earned the rank of elder-statesman, and perhaps the right to be venerated as a wise elder. Many at this stage of life, while still being actively involved in family support, will volunteer their services to the community, either for a local cause or perhaps for the advancement of national or international causes or societies. This is the season for creation and contemplation, and as Thompson notes, many geniuses are at their best at this stage. Optimists will tell us that contrary to the popular adage that 'life begins at 40 or 50', it really begins at 60! We can credit icons like Tina Turner and Stevie Nicks for helping create the new 60, which, if nothing else, reminds us that as life expectancy increases, consciousness is necessarily evolving new complementary structures.[24]

Spiritual emergence

As suggested above, despite dynamic and progressive or cyclic shifts in consciousness identified by Piaget and the *ludens-faber* model, most such shifts can be said to be more or less gradual when measured on day-to-day time scales. Some shifts however are more abrupt, and more fundamental. The shift from simple to self-consciousness which takes place between one and two years old (between Piaget's first and second stage) is momentous to say the least. The gurgling, flailing 'infant' suddenly becomes a mobile young person able to enter and navigate the communicative, linguistic world of the adult. In most people's estimation, this process of becoming self-conscious is what makes us truly human.

Several students of the evolution of consciousness have argued that our earliest ancestors were not self-conscious. In other words they lived in what, by our standards, would be a dreaming or sleeping state. We shall examine these propositions soon, but first we must return to the notion of the sudden epiphany or religious experience that seems to be a species of mid-life crisis or transformation. According to many consciousness and mysticism traditions, such epiphanies lead to permanent shifts in consciousness. In his classic 1901 book *Cosmic Consciousness*, Richard Bucke, a renowned Canadian psychiatrist, claimed to have documented several dozen cases of abrupt shifts to 'higher' cosmic consciousness using clinical and historical records.[25] He documented 43 cases where such transformations took place more or less abruptly and calculated an average age of 37 years for 34 of these individuals whose age was known. He also reported that he himself had had the same experience at age 35, and claimed that the transformation most often took place in the springtime. He also held that all the subjects he studied had shown a profound shift towards enhanced intellectual and moral sensibilities. Most could be described as geniuses, creative artists or spiritual gurus.

Although few people have conducted similar surveys in the intervening century since *Cosmic Consciousness* was published, those that are available are consistent with Bucke's conclusions. One of us (M.L.) used various sources on 'sudden' spiritual emergence to double Bucke's sample size, to about one hundred persons, still showing that the average age at which the cosmic consciousness transformation took place remains very close to 37 years of age, and is not significantly different for either men or women.[26] Subsequently we found another study, by Basil Douglas-Smith, based on 211 individuals, which clearly shows that the peak, at aged 35, is very pronounced.[27] Smith, however, also identified as prominent a peak at age nineteen. However these results are interpreted, and we have combined them into a single chart (see Figure 3.3), it is clear that three independent studies confirm the tendency for spiritual epiphany in the mid-thirties. Likewise the identification of an earlier peak at around age nineteen suggests a rhythmic pattern to spiritual life tied in some way to our basic developmental biology, and perhaps in this case to our 'coming of age'.

Bucke was a contemporary of William James, a distinguished Harvard psychologist and brother of the famous novelist Henry James. William authored the well known book *The Varieties of Religious Experience*, published in 1907, which delivered much the same message as *Cosmic Consciousness*.[28] It is interesting to note that both Bucke and James took the phenomenon of sudden epiphany very seriously and considered it a legitimate subject for serious study in the field of psychology, where both were towering figures in their day.

Beginning in the 1960s, western interest in consciousness began to frame discussion of life-changing shifts in consciousness in a new light, born of a very old tradition, which had already begun to seep into western culture in the late 1800s.[29] In his book *Kundalini: the evolutionary energy in man*, Gopi Krishna spoke of his sudden spiritual emergence as the experience of the rising of the *kundalini*. The kundalini, meaning 'coiled serpent' in Hindu traditions dating back to the Axial Age (~500 BC), refers specifically to energy that resides at the base of the spine, in the lower chakras, like a coiled serpent.[30] When released, this energy rises up the spine illuminating the higher chakras, and leaving those who experience it with a profound and enduring sense of a powerful transformation in consciousness (see Chapter 9). This release can be quite spontaneous. It can result from years of meditation, as in the case of Krishna's experience, simply as the result of a shock to the base of the spine, or for no obvious reason. In this sense it is a sudden developmental and 'cosmic' shift in consciousness much as Bucke outlined.

Following Krishna's autobiographical accounts, various western physicians, psychologists and students of consciousness like Basil Douglas-Smith, Isaac Bentov, Lee Sannella and Jean Galbraith, have reported other such kundalini cases, most of which were welcomed and regarded as beneficial spiritual transformations by those who reported the experiences.[31] Although it is sad to report that such cases are still not well understood in the West, it is becoming increasingly clear that our lack of understanding stems largely from ignorance. In some cultures such shifts in consciousness would be cultivated as a shamanic gift and regarded as desirable. The Oxford biologist Alister Hardy took a profound interest in such spiritual experiences and in later life began to collect hundreds of case histories for his Religious Experiences Research Centre.[32] It soon became clear that such experiences were not so much rare as rarely reported. Interestingly, however, the number of cases being reported to the RERC, under the label of 'kundalini

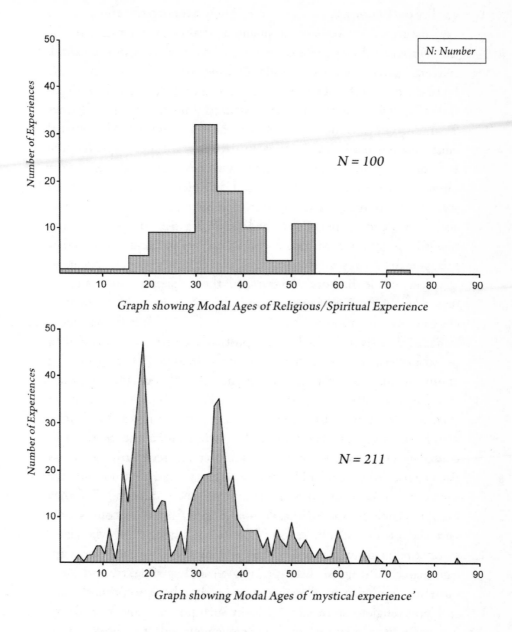

Figure 3.3. Charts showing tendency for spiritual epiphanies to be experienced at around 19 and 35 years of age. (Lower histogram from Douglas-Smith 1971. Upper histogram after Bucke 1901, Lockley 2000 and other sources.)

awakening', continues to increase. Such experiences are by their very nature subjective personal phenomena akin to those reported by the mystics, and it is possible to refer to them as religious, mystical, spiritual, psychological or kundalini experiences. In the latter case however, powerful forces and physical sensations associated with an intense energizing of the nervous system is unmistakable (see Chapter 9). Thus, we can conclude that the broader umbrella of spiritual emergence is a manifestation of perennial 'self-realization' aspects of the evolution of consciousness associated with developmental stages most often reported in early adulthood and mid-life. Though they may be spontaneous in many cases, suggesting an intrinsic biological dynamic, they may be brought on by meditative practice, or perhaps even unwittingly by right attitude and unconscious preparation. Whether the epiphany is consciously sought, desired or anticipated, it appears to be a general rule that, once experienced, the recipient is caught up in a new 'adventure of consciousness' and very much aware of the shift in consciousness structure.

Thus, there is little doubt that spiritual experience is 'real', and in no obvious way related to delusion or hallucination. (There is a strong argument that all subjective experience is real, but that objective interpretation of such experience is another matter.) One might say that, by definition, spiritual emergence experience is real for anyone who honestly reports such profoundly moving and authentic shifts in consciousness and their psychological impact. Alister Hardy's work has been continued by David Hay and others, who note that not only are such experiences common, and often intense and sudden, but that they occur in people who are ostensibly 'non-religious' but often comfortable with the broader label of 'spiritual', as well as among those who claim distinct religious affiliations.[33] Thus, as Hay notes, we can if we wish, legitimately distinguish between religion and spirituality, and align ourselves with what we already referred to in Thompson's vocabulary as a 'post-religious spirituality'. Clearly such post-religious spirituality is both an individual and a collective phenomenon. It is the collective aspects of a broad range of consciousness phenomena that we shall examine in the next chapter.

4. Ebb, Flow and Cycles in Collective Consciousness

Changing generations: the fourth turning

Life is a cycle. Or, as the combination of daily life and the longer-term *ludens-faber* paradigm suggests, a complex of cycles within cycles. This is our common experience: the daily cycle of waking and sleeping is embedded in the lunar cycle which is in turn embedded in the annual cycle when nature alternately sleeps and awakens. The yearly cycle is embedded in the seven year cycle and much larger cosmic cycles such as the precession of the equinoxes (25,920 years) which results from the cyclic shift in the angle of the Earth's rotational axis. What concerns us here is the influence that such natural cycles have on human consciousness and behaviour throughout longer historical periods.[1] We have already seen that the seven year cycle is very significant in human development, and should be taken seriously as an organic guide to education, maturation and shifts from childhood to adult and mature, sage or cosmic consciousness.

Such cycles therefore seem to make good organic sense, and we may ask how the human life cycle is embedded in history, giving us generations with different collective experience. The American historians William Strauss and Neil Howe have devoted much research to the dynamic shifts that accompany the sequence of generations in western society, giving us what we sometimes refer to as the cycles of history. In their book *The Fourth Turning*, they suggest that society undergoes a four-generation cycle of between about 85 and 100 years.[2] This cycle, or *saeculum* in Latin, from which we get our word secular (related to time or age), is by no means a new idea. We have already seen that we can trace this cycle back to early Hindu philosophy and the Ashrama tradition, which breaks the individual human life cycle into four phases. We must remember that the number 21 is significant, and one wonders if the West might not have inherited numerical ideas from ancient eastern traditions. The number 21 shows up in the Ashrama tradition as well as in the biological *ludens-faber* cycle, in our legal system and even in currency (the guinea is 21 shillings). It is 4 x 21 which approximates the *saeculum*. The number four is significant in many cosmologies, including our own. There are four seasons, and four cardinal directions. These seasons are very precisely circumscribed with two solstices and two equinoxes.[3] Around 1000 BC the Brahmana commentaries on the Vedas made a detailed analysis of cosmic time, or deep time as geologists and astronomers sometimes call it. In this system, the *kalpa,* which symbolically represents a single day in the life of God, is 4.32 billion years in duration. This number is remarkably close to twentieth century estimates of 4.6 billion years for the age of the Earth and solar system. A smaller division of time, the *maha yuga,* is only 4.32 million years (coincidentally about the age of our hominid family). The *maha yuga* is divided as follows into four phases or *yugas.* (See Table 4.1.)

Table 4.1. *The early Hindu conception of time predates the schemes of western science by several millennia.*

Hindu time or *yuga*	Duration in years	Human experience
krta yuga = golden age	1,728,000	100% health/very long life
treta yuga = silver age	1,296,000	75% health/long life
dvapara yuga = bronze age	864,000	50% health/medium life
kali yuga = iron age	432, 000	25% health/short life
	maha yuga total 4,320,000	

What is of interest here, beside the early emergence of a type of quantitative or scientific time consciousness in India, is the use of a four stage cycle with an arithmetic progression in which the golden, silver, bronze and iron ages occupy an ever-diminishing 'slice' of time, from 40% to 30, 20 and 10% respectively.

By comparison, in the Strauss-Howe, Fourth Turning model, the four sequential generations of the century (*saeculum*) are described as spring, summer, autumn and winter. They argue that these are the successive or cyclic moods of society, equivalent to childhood, young adulthood, middle and old age. Although they do not say so explicitly, these phases parallel the Ashrama model. Based on the modern, post-Renaissance history of the West (Europe and North America), they argue that one can trace a series of crises (winters) which erupt every 85–100 years. These are followed by renewals (spring), and the stabilizing of society, restless awakenings and ferment (summer) and unravellings (autumn) followed by the next crisis.

As shown in the accompanying chart (Table 4.2), from the War of American Independence (1776) up to the time of the Ronald Reagan years, Strauss and Howe identify thirteen American generations. This explains the popular term generation X or 'thirteeners'. The War of Independence coincided with crisis in Europe: notably the French Revolution and subsequent Napoleonic wars. This was also the time of the Industrial Revolution and much attendant upheaval masterfully described by Jacob Bronowski in *William Blake and the Age of Revolution*.[4] Four generations later, America erupted in civil war, and four generations after that came Pearl Harbour and US involvement in World War II. Now in the 2000s decade, the country is involved in an epic ideological battle labelled the 'war on terror'. Although these times represent bad news for those wishing to live in peace and harmony, history tells us that peace should come after the crisis generation, and society will reconstruct itself, as it did when creating a new nation in 1776 or when rebuilding after the civil war and after World War II. The baby boomer generation for example is tangible proof of biological rebirth. So good news follows bad just as bad times eventually follow good times.

Table 4.2. A simplified summary of the Fourth Turning cycle of generations model proposed by the historians Strauss and Howe. Time sequence reads up from bottom row, and from left to right

Spring rebirth	Summer awakening	Autumn unravelling	Winter crisis
1945–63: birth of baby boomers>	1963–80: civil rights and consciousness >	1980s: Reagan years, gang & culture wars >	2000 decade: 9/11 terrorism >
Reconstruction>	1880–1900: women's vote; new psychology >	1900–20: roaring 20s; World War I; gangsters >	1929–45: Depression; World War II >
1789: Washington's Bill of rights >	1820s: Romanticism; Abolitionist debates >	1840s: Mexican-American war >	1860s:Civil war >
		1750–76: colonial unrest; land grabs >	1770s–1800: War of US Independence >

But as society rebuilds itself after a crisis, and settles down to a conventional, even a complacent peace, it is not long before a new generation comes along with unconventional demands for new rights. So the new society is tested by the awakenings of new moral conscience and new ideals. After independence it was the abolition of slavery, then after the civil war it was the emancipation of women, followed, four generations later in the 1960s, by civil rights and the consciousness generation. This idealistic ferment has been characterized as the summer of spiritual awakening, but it can also leads to excess as the demand for individual rights gives way to greed and selfishness. So in the unravelling generation we see society become divided by culture wars, corruption and increased crime and social unrest. This leads to mass movements and migrations as were seen in the 1760s, again in the 1840s, the early 1900s and the 1990s. It is no coincidence that gang warfare erupted in the streets in two successive periods of unravelling (the 1920s and 1990s).

Strauss and Howe go into more detail about how each phase of the cycle has its own mood which in turn spawns its own distinctive birth generation. The spring produces 'prophets' with an optimistic vision of the future, who come of age in the summer awakening. Summer produces 'nomads', restless offspring of the visionaries, the 'latch key kids' who have to deal with society's unravelling when they come of age. Autumn produces 'heroes' who have to mend society when they become young adults during the crisis, and winter produces 'artists', the children of the heroes, who have to be protected and fall back on their own inner resources while the crisis plays out.

The Fourth Turning cycle also oscillates between emphasis on community coherence during the crisis and spring reconstruction phases, and expression of a more self-indulgent individuality during the summer awakening and the autumn unravelling. This interesting polarity leads to many of the 'alternation of generation' dynamics of our epic dramas. In seasons of awakening the young prophet delivers a spiritual message to the corrupt rulers of the old empire — as Jesus did to Herod, or as Martin Luther King did to materialist, consumer-conscious America. Or, in crisis seasons, the roles may be reversed as the old prophet takes the young king as his apprentice to train him to rebuild secular society, as Merlin did with Arthur, or as Obi Wan Kenobi does with Luke Skywalker. These alternation of generation dynamics remind us, at least symbolically, of the polarities and resonances of solstice and equinox.

History's masculine and feminine moods

As we examine the pattern of cycles within cycles we soon see that the seven year cycle helps define the 21 year generational cycle which in turn defines the saeculum, and half saeculum alternation of generations. If we examine the century-long cycles of history, even during the short history of modern science from the Renaissance to the present, we cannot fail to see the swing of the pendulum between emotional and intellectual poles (see Table 4.3). With the rise of industrial society, impersonal technology, military and political power, we see the inevitable reaction or backlash, and a strong yearning for reintegration with organic nature. Historians more or less name the even-numbered centuries (sixteenth, eighteenth and twentieth) as those in which intellectual, scientific and technological pursuits shaped the evolution of western culture. Conversely the odd-numbered centuries (seventeenth, nineteenth and twenty-first) mark periods of greater emotional and artistic ascendancy.

Table 4.3. The low frequency rhythm of history: century-long intellectual and emotional cycles

Renaissance	16th century intellectual ascendancy.	New mechanisms
Baroque	17th century emotional ascendancy.	New Colonialism
Enlightenment	18th century intellectual ascendancy.	Industrial Revolution
Romanticism	19th century emotional ascendancy.	Romanticism
Technology	20th century intellectual ascendancy.	Modernism
New Age	21st century emotional ascendancy.	Post-modernism

According to Karen Armstrong, during the Axial period, which she describes as *the Age of Transformation*, each century from 900–200 BC marked a discernible advance in moral and ethical evolution in some sphere of consciousness.[5] She even lists eight sequential domains dealing with: 'ritual', 'kenosis', (meaning an emptying of self, or dismantling of Ego) 'knowledge', 'suffering', 'empathy', 'concern for everybody', 'all is one' and 'empire', which preoccupied history and consciousness during successive centuries during this period.

We can ask whether this cycling between emotional and intellectual intelligence represents some sort of inherent polarity in the human condition. Is there, one wonders, any relationship to gender as in Yin (feminine) and Yang (masculine)? Robert Keck suggested that human history can be divided into three major epochs:

—Epoch I is characterized as our infancy, when we lived in the bosom of nature, with our feminine sensitivities more prominent. This was the Paleolithic and early Neolithic epoch of the Mother Goddess, when feminine mystery, wisdom and fertility were revered.

—Epoch II represents our adolescence, when masculine aggression began to gain ascendancy over feminine sensibilities, and we entered into the empire building and warring phase of history which brought us from the first cities to today's megalopolis. This epoch sees the development of identity and the individuation of Ego.

—Epoch III represents our maturity, which we have yet to reach, though the positive signs suggest the new age is dawning, and Keck claims it is as inevitable as any other long-term evolutionary development. It anticipates a return to the feminine, and the integration of masculine and feminine principles. Keck describes it as the epoch of the butterfly emerging into the light of higher consciousness from the dark obscurity of the wormlike chrysalis.[6]

This three-epoch model is reminiscent of Riane Eisler's exposition on human history in *The Chalice and the Blade,* in which she contrasts the two separate halves of history as the feminine (Goddess) and masculine (Androcratic) epochs and speaks of their potential for fusion in an integrated 'Gylany' partnership.[7] Just as Epoch I represents the time when our simple Id consciousness was embedded in nature, and Epoch II represents the self-centred, Ego consciousness, so Epoch III represents a mature, sage or Superego consciousness. *Psyche*, the Greek word for butterfly, symbolizes the Superego and the liberation *(moksha)* of higher consciousness. Here we see that a cyclic model has taken on a decidedly progressive overprint. This is in fact a consistent theme in all the schemes we have examined, and there should be no contradiction in the conceptual integration of cyclic and linear time. As we shall see, they become integrated in the spiral dynamics model.

Slowly coming to our senses

Students of the humanities often comment on historical cycles in the moods and fashions of literature and the arts. Such work is influenced by mood swings in the collective unconscious, and its manifestations in power politics. In his book *Coming to our Senses,* Morris Berman writes that during western history we have tended to lose touch with our physical and emotional experience as our rational minds have developed.[8] However, periodically our emotional bodies rise to the surface to disturb our fragile mental equilibrium and reshuffle our ideologies. These historic cycles of thought and feeling are much like the rather simple century-long cycles just outlined. They can be seen as a type of *faber-ludens* oscillation on an extended time scale. As the pendulum has swung back and forth between these two poles of human experience, various religious and ideological movements have attempted to re-connect with physical (somatic) experience, so as to escape the sterility of abstract doctrine. Often this is accomplished through the eruption of heretical movements that fly in the face of orthodoxy, which by its very nature tends to rely on a more rigid ideological doctrine for its authority.

The 'particular problem in the history of Western Civilization', writes Berman, is 'the presence of a heretical or countercultural tradition that is rooted in bodily experience and that rejects the cerebral, or formulistic, way of life of the dominant culture (orthodoxy)'. In many ways the 'story of Western Heresy is that of perpetual refuelling from Eastern influences'. This leads to the overturning of the masculine mentality by more ancestral or original feminine traditions. Moreover, these patterns of behaviour 'most likely have an underlying somatic basis' leading to 'a kind of independent and spontaneous recovery of information'.[9] Here we note that the polarity between the archaic, 'original' East and modern West is seen as analogous to the polarity between feminine and masculine. This suggests the notion that in culture and human affairs things begin in an innocent pure state and become complex and corrupted: or as Charles Péguy put it: 'Everything begins in mysticism and ends in politics.'[10] Jean Gebser put it another way, saying that things begin efficiently and end deficiently (see Chapter 5).

Perhaps it is hard not to put value judgments on the relative merits of cultural dynamics as they swing from intellectual to emotional ascendancy and back, because in real life such shifts from efficient to deficient modes may cause genuine concern. Berman's observations demonstrate that these oscillations are clearly complex. In general the historical cycles so far outlined are readily recognizable, but in any cultural place and time the dynamics may be complex and hard to decipher. Simple cyclic patterns may overprint one another and when coupled with abrupt events, the ups and downs of history may appear 'messy' and difficult to pigeonhole in convenient schemes. However this does not mean that the cyclic march of history is a purely messy and chaotic random walk. We may aspire to see order and pattern arising from the chaos.

Berman begins with the example of animism, the ancestral feeling or experience that Owen Barfield and others describe as 'participation' in the vital living cosmos.[11] Thus, human conscious experience was originally deeply and unconsciously embedded in somatic, physical, spiritual or mystical experience. We shall attempt to excavate this archaic consciousness structure in the next chapter, but in the meantime let us review Berman's thought.

Traditionally the western mindset reveres the Axial Age (~500 BC) for bringing us recorded history and the intellectual traditions of reason and rationalism, which we sometimes refer to as the dawning of thought or more appropriately 'abstract thought'. Although this axial 'step forward' into history helped the human mind emerge from archaic animism, this intellectual phase of development was deeply embedded in mythological experience. So, with Greek culture we see interesting contrasts between a transcendental idealism and a new empiricism: that is, Platonic idealism on the one hand, and Aristotelian empiricism on the other. This was an early western manifestation of ensuing dualistic debates on the perceived split between body and soul (or spirit).

Discussion of perceptual and conceptual differences between the rational, sense-accessible or tangible world and the intangible experiential world could easily detain us for many chapters. In the context of Berman's ideas on western history, it suffices to contrast the so-called *Gnostic* tradition (referring to what can be characterized as direct, individual experiential 'knowing' of the divine or intangible) and collective, institutionalized forms of 'religion' which focus on the tangible institutionalized symbols, idols and doctrines of a well defined canon. The Gnostic tradition, which arose in the first century AD, can be considered a type of mysticism akin to Bucke's cosmic consciousness, which by its very nature resists scholastic intellectual analysis and precise rational definition. Nevertheless, despite the influence on western philosophy of notable mystics like Plotinus (AD ~204–270), they still represented a minority. As a result throughout most of western European history, the mystical impulse has always been overtaken by the institutionalizing forces of organized religion (see Chapter 8). Most famously this happened at the Council of Nicea in AD 325 when Constantine made Christianity the official Roman state religion.[12] Again, 'everything begins in mysticism and ends in politics'.

Traditional histories of western civilization speak of a dark age that endured from the fall of the Roman Empire until the medieval 'Renewal' and subsequent 'Renaissance'. This is not strictly true and it is certainly a Christian-biased view. Not only did mysticism enjoy moments of glory thanks to Plotinus and others, but the spectacular rise of Islam in the seventh century, and the occupation of Spain by the Moors, fostered a highly sophisticated, pluralistic and scholastic civilization that endured in Andalusia for centuries. This period between the eighth and thirteenth centuries became known as the Golden Age, and the cities like Cordoba, Sevilla and Granada became Andalusian jewels in 'The Ornament of the World' as pilgrims crossed Europe on foot just to marvel at the cultural splendour.

The Andalusian influence spread across Europe, and by the twelfth century, sometimes referred to as a pre-Renaissance age of Renewal, scholasticism was on the rise. Portraits indicate an increase in self-awareness. Interiority — the kinesthetic self — was emerging from medieval slumber. The appearance of courtly love, troubadours and the shift towards more realism in art are all very apparent. The Madonna becomes a sensuous woman, not an icon. On the fringes of the Andalusian 'empire', the little French village of Montaillou just north of the Pyrenees provides a fascinating insight into life between 1294 and 1324.[13] Montaillou became a centre for the Cathars (or Albigenses) a progressive religious group probably with origins in far off Persia. By the eleventh century, the culture became established in the Languedoc region around Toulouse, France, and despite claiming to be Christians, they were treated as heretics by the Roman Catholic Church, and as many as 500,000 were mercilessly massacred by Pope Innocent III from 1208 onwards. (How one wonders did the name Innocent ever stick in such circumstances!) Montaillou became infamous as the site of a protracted inquisition by the Church authorities into Cathar spirituality. This, Berman asserts, shows the profound differences between the experiential flavour of Cathar aspirations and the rigid authoritarian doctrines of Roman Catholicism imposed at the time.

At this time, gifted scholars like Roger Bacon were in danger of persecution as heretics. Mystics such as Nicholas of Basle (c.1308–97) of the free-spirited Beghard community, with connections to the Cathar tradition, paid the ultimate price for their unorthodox persuasions by being burned at the stake. These people often held the heretical or 'Gnostic' view that through mystical experience one may glimpse a reciprocal personal relationship with God, or as Meister Eckhart famously put it: 'The eyes in which I see God are the same in which he sees me'.[14] Although some like Hildegard of Bingen (1098–1179), reported visionary connections with, and guidance from, God, without being charged with serious heresy, they still walked a fine political line. Others like the celebrated mystic Teresa of Avila (1515–82) also faced scrutiny from Church authorities for daring, like Eckhart, to speak of communing directly with the divine. As women, their mystical inclinations were doubly problematic as they were generally forbidden from expounding on theology or directly affirming a woman's spirituality.

What was really at stake (note the metaphor) in these drawn-out centuries of Renewal and early-Renaissance transformation was a battle for the individual soul — what William Irwin Thompson calls 'the fall of the soul into time, the entrapment of an angelic soul in the body ...'[15] Did one dare to profess an interiority sufficient to commune with, let alone explain or understand the workings of ones individual soul? If so, was one not usurping the role of God, and his supposedly divinely appointed ministers on earth, in prescribing what was right or wrong? Dared one be in charge of an interior dialogue that could lead to one's own salvation and redemption, without making up one's own rules. Most of the thousands of pages of inquisitorial transcripts from Montaillou speak to an earnest exploration of these soul questions — matters that today might be labelled personal conviction. As we shall see in the next chapter, this was all a fundamental, and at the time very dangerous, part of the process of individuation.

Catharism's inclination towards the exploration of personal, interior experience was perhaps a foretaste of the Italian Renaissance which broke open medieval church scholasticism and helped transform medieval magic into modern science, but not without inevitable struggles against entrenched orthodoxy. Copernicus (1473–1543) saw science through the eyes of a mystic and regarded nature as the path to God.[16] Giordano Bruno (1548–1600) was executed by the Inquisition for his attempts to reconcile science and religion. For example, he believed that a scientific Copernican universe might be part of God's system of infinite worlds.[17] Music was linked to astronomy and the 'music of the spheres' became a metaphor for the harmony of the planets. After all, as the Pythagoreans knew long ago, octaves provided mysterious evidence that the subtle organization of the universe resonated with stirring, mathematically harmonious vibrations that manifest in the sensible world. Nevertheless, despite the grip of the Church and the conviction of most late medieval and Renaissance scientists that natural law was divinely inspired, science slowly drifted away from the experiential and alchemical study of the ascent of the soul, and became theoretical and representational — more like what we know today.[18] Likewise music became popular and a force for social cohesion and uniformity linked to the idea of mathematical law and order (an imaginary system divorced from experience).

Berman ends his exposition on these cycles of history with his take on the rise of Nazism as an unconscious expression of a deep emotional eruption in the psyche. Hitler was a pro-vegetarian, anti-smoking health fanatic. He was interested in the Knight Templars, the search for the Ark of the Covenant and the Holy Grail and had strange delusions about a hollow earth.[19] He derived some of his more esoteric ideas about the twisted cross (swastika) from the occult writings of the theosophist Helena Blavatsky who had ideas about a higher race: the Aryans (but not in the sense of the linguistic culture going back five thousand years).[20] National Socialism was a type of secular Gnosis according to some historians.

Hitler and Goebbels had Road-to-Damascus type 'divine message' experiences, except that in Hitler's case his 'inner rapture' — his supernatural vision — was evidently due to the noxious influence of World War I mustard gas! They believed in their experiences and their redemption. In perpetrating the slaughter of Jews — to purify German blood — by spilling that which was impure, Hitler was re-enacting a perverse form of archaic sacrifice. His experience has been labelled a 'dynamic fallacy' connoting the misleading idea that all psychic energy and epiphany is divine or positive. At a Nuremberg rally in 1937, the mayor of Hamburg claimed that they needed no priests or parsons because they could communicate directly with God through Adolf Hitler. Others called him an 'enchanter' who channelled his oratorical words hypnotically from some kind of magic association with the universe.

Pattern and structure in revolution and renaissance

From Strauss and Howe, to Berman and Keck, many have championed the idea that history reveals patterns of cyclic ebb and flow, and enough has been written to suggest that we can define these patterns, at least loosely, and relate them to emotional, intellectual and political currents to which we assign labels such as age of Romanticism, Neoclassicism, Renewal, Renaissance, and so forth. We may not be able to define these large scale patterns very precisely, and we may find pattern embedded within pattern, but in stating such classic aphorisms as 'those who do not learn the lessons of history are condemned to repeat them',[21] it is clear that we recognize that such cycles are deeply embedded in what we casually call the historical past. These ages or chapters in history may be subtle or obvious, they may develop gradually, or be precipitated by sudden, revolutionary events. To add to our growing list of mostly western examples, we can add the observations of the Chinese Scholar Lin Yutang, who claimed that Chinese civilization had gone through three phases of unprecedented development. In his book *My Country and My People*, written in 1924, he identified three peaks of development occurring about eight hundred years apart.[22] (See Figure 4.1.) The last, which began with the rise of the Ming Dynasty in the fourteenth century, ostensibly predicted another rapid rise by the twenty-second century. More recent history rather clearly shows that the fourth peak came a little early, around the dawn of the twenty-first century.

These collective, historical cycles, and their subtle emotional-intellectual oscillations are summarized in Table 4.4. Although somewhat simplified, they nevertheless reflect a synthesis of the observations of many competent scholars, and will no doubt provide fertile soil for further debate.

Table 4.4. Historical cycles and their approximate duration, compiled from various sources. Note that the short wake-sleep and faber-ludens cycles have well defined biological correlates.

Collective, Historical Societal-Cultural Cycles	Duration
Wake-sleep/*Faber-ludens* phases	24 hours–7 years
Strauss-Howe generations	21–25 years
Fourth turning *saeculum* (century)	~85–100 years
Berman's rational-emotional cycles	~100– ~500 years
Lin Yutang's cycles	~800 years
Keck's epochs	~5,000–10,000 years

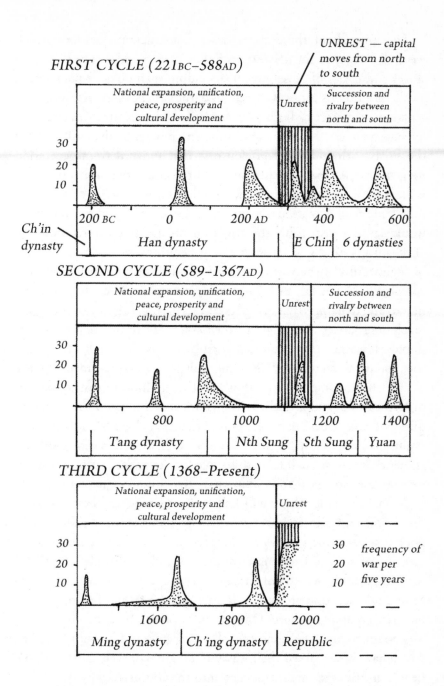

Figure 4.1. Zeniths of development in Chinese civilization. (Recompiled after Lin Yutang 1935.)

Thinking about these cycles raises fundamental questions which we shall carry into the discussions in Chapters 5 and 6. The biological stages and cycles of development, including such obvious phenomena as the 24 hour sleep-wake cycle and the seven year cycle of tooth replacement, affect us both as individuals and collectively as a species. So one wonders, is there some inherent biological dynamic behind the shifts in mood and consciousness that proceed from generation to generation, from century to century, and from millennium to millennium. Although the answer could be yes, if we were to invoke some inherent evolutionary dynamic, as yet not understood, most would say that ostensibly the answer is no! Despite the reproductive 'mechanisms' that give us successive generations, we have no obvious biological correlate between the mood of a spring or summer generation compared with a fall or winter generation. Likewise, the changing moods and dynamics that bring about ages of romanticism, or industrial revolutions, are not usually explained as a result of inherent biological impulses. Rather they are usually explained as the result of external socio-political factors (or in some cultures, by bad omens in the weather or alignment of the stars). So the question is: can we deduce different causes and influences for historical cycles that operate on very different time scales? Bearing in mind that there have been remarkable evolutionary cycles, spanning billions of years, as well as geological and climatic cycles, we shall leave this topic for now and return to it when we examine biological and evolutionary dynamics in Chapter 6

A final example which will help point us towards these scientific discussions in Chapter 5, was provided by Thomas Kuhn in his classic book *The Structure of Scientific Revolutions*.[23] As his title suggests the progress of science is not just the result of clever people coming up with inspirations which enlighten others and extract new truths from analysis of the world around us. Rather it is the result of a process that has pattern and structure. One of Kuhn's more useful contributions was to introduce the concepts of paradigms and paradigm shifts, which have now worked their way into everyday speech. By definition, new scientific ideas are introduced into the community by handful of innovative pioneers. The rest of the scientific community is usually initially reluctant to accept such ideas. There are many reasons for this, including genuine disbelief, a training which weds one to old paradigms, and other less justifiable reasons such as ignorance and professional jealousy. So the innovator usually has to go through a three-stage process in which he/she is: (1) ignored; (2) reviled and criticized; and finally (3) accepted and even revered. None of this seems like the playing out of a rational and objective analytical process in which all participating scientists give equal time and thought to hearing and analyzing the old and new hypotheses. So the process is often, messy, intense, drawn-out and emotionally-charged, rather than cool, calm and collected. In short, for historical reasons the old ideas, have mass and inertia as established paradigms, whereas the new hypotheses are untested, risky and yet to achieve paradigm status, regardless of how brilliant and fruitful they may subsequently prove to be.

The physicist Stephen Edelglass has reminded us that Isaac Newton (1643–1727), regarded as a giant of modern science, was also deeply interested in alchemy, which today we too often misinterpret superficially as pseudo-scientific mumbo jumbo. In fact Newton's alchemical experimentation was, as already intimated, an attempt to understand the spiritual or cosmic forces that reside within us. Indeed in his theory of gravitation, Newton proposed centric forces, within bodies, that symbolize independence and individuality (rather than the external Copernican forces that keep the planets spinning in space). It is probably no coincidence that this scientific revolution coincided with the aforementioned increase in use of self-aware and self-conscious vocabulary celebrating the new sense of 'Enlightenment' identity.[24]

We have no doubt that however, rational and even-handed we try to be in the exposition of our hypotheses, even those that, in our opinion, are already achieving 'new paradigm status' there will be those who will register both intellectually and emotionally charged objections. Even more likely there will be those who want to explain, test or otherwise 're-structure' various hypotheses by importing them under the old umbrella of various favoured, statistical, sociological, economic, genetic, physiological and psychological paradigms. All well and good we say, but our aim is to keep an eye on the broadest possible vistas. The narrow paradigm of the trees may compromise the possibility that beyond lies a richer woodland landscape.

Time and space consciousness: before and after memory

So far all of our explanations for stages and cycles of development in individuals and during the broad sweep of history have come from the pens of modern scholars. This at least lends consistency to our thinking, but it has given us limited insight into how our ancestors perceived the world. How has space-time consciousness changed over the centuries and millennia? As we shall see in Chapter 5, a few scholars have looked into this question, with fruitful results. But, as a prelude to these in-depth discussions we will review the conventional wisdom, and the implications of what we moderns like to call some relatively well established 'facts'.

Leaving aside certain seemingly remarkable proto-scientific insights about the age of the universe from remarkable Vedic and Hindu sources, most histories of western science, only take us back to the Axial Age in ancient Greece some 2500 years ago. At this time we may ask what the concepts of space and time meant to even the most inquisitive and intelligent individuals. Did they, as Santillana and Dechend suggest, understand the complex astronomical cycles like the precession of the equinoxes? The conventional way such questions have been answered is by interpreting ancient inventions and artifacts, and written records. We tend to select these because they deal most directly with items relevant to our concepts of space and time today. They include rulers, maps and clocks that measure and record the 'dimensions' of space and time. Among the Pythagoreans geometry was just being invented, and although Aristarchus knew the world as a sphere revolving around the sun, he was an exception, and many in the intervening 1800 years, prior to Copernicus, believed the world to be at the centre of the solar system. Likewise conventional wisdom holds that, despite the use of sundials, candles and hour glasses we did not invent reliable mechanical clocks, understand longitude or draw relatively accurate maps until well into the modern scientific era: that is, the sixteenth and seventeenth centuries. It is well known among historians that a variety of confusing and conflicting calendars were used before the Gregorian calendar (named after Pope Gregory III) was introduced in 1582 to replace the

Roman or Julian calendar (named after Julius Caesar). In comparison with the very slight, split-second adjustments made to atomic clocks today, and the difference in global time zones, pre-1582 calendars could differ from one another by several days.

We probably have a rather biased, modern view of accuracy, accentuating pride in our own contemporary technology. But in the absence of a complete historical record, it is hard to really know what our ancestors knew or cared to know. There is intriguing evidence that some among the ancients were very good astronomers, and map makers. For example, authentic maps from 1513, or earlier, suggest that somebody may have mapped the Antarctic in significant detail three centuries before its supposed 'discovery' by western explorers in 1818. However, according to others, the maps may also be crude depictions of Australia.[25] The Antarctic map hypothesis suggests that ancient, pre-classical, civilizations developed knowledge that western civilization has since lost. As the Aristarchus example shows this is clearly possible. But lost is lost, until rediscovered. Even if we admit, despite our worship of progress, that our consciousness may have regressed in some areas and lost touch with some ancient wisdom, we can nevertheless excavate what is still preserved. Thus, leaving aside speculation about the potential sophistication of pre-Axial Age civilizations, we can still look at how ancient cultures perceived space and time using the record of the last 2,500–3,000 years.

In *The Discovery of Time,* the authors, Stephen Toulmin and June Goodfield, quote R.G. Collinwood as saying: 'Nature has no history. Only human beings have history'.[26] This inference is perhaps self-evident in as much as animals have created no records of past activities. It is historians who invented the word history and human paleontologists who have reconstructed animal pre-history. As Toulmin and Goodfield note: 'Each of us has one immediate direct link to the past — memory'.[27] The human-animal distinction also echoes the rather more subtle and esoteric ideas of Rudolf Steiner. In his book *Cosmic Memory,* he describes the various stages in the development of memory by linking them to structures of consciousness to which he assigns various 'race' or cultural labels.[28] In simplified, tabulated form, his scheme is shown in Table 4.5.

Table 4.5. A brief summary of the evolution of human memory in the framework of Rudolf Steiner's esoteric cultural scheme.

Rmoahals: this race (or consciousness) manifested the first feelings of memory attachment to vivid sense impressions. There was a bonding between the soul and things outside man. (Ostensibly a very early stage of self-consciousness.) Memory was connected to language, so words were therefore sacred and powerful.

Tlavatli: this race (or consciousness), as a result of memory, began to feel its own value and develop ambition. Individuals wished to receive recognition from the community for deeds remembered. This led to the foundation of ancestor cults. Those who did great deeds might be regarded by others as masters of natural forces.

Toltecs: this race (or consciousness) founded community consciousness, the seeds of which were sown earlier. As memory developed deeds of ancestors were not to be forgotten by descendants. Personal experience became important, and led to the founding of new communities, trying new things. Powerful persons were 'initiated' into eternal laws of spiritual development.

Primitive Toranians: this race (or consciousness) was instructed in power and mastery over life forces, but became selfish and exploited their fellows.

Primal Semites: this race (or consciousness) developed logical thinking, which restrained the excesses of power. Past experiences were compared and individuals could check appetites. Impulse for selfish action transferred within, but inner control meant loss of external control of nature.

Akkadians: this race (or consciousness) developed thought more. Role of leadership conferred on intelligent persons not just one ones whose dramatic deeds were remembered. A fondness for innovation, commerce, colonization helped formulate laws, and regain loss of control over nature.

Mongols: this race (or consciousness) is thought to have been highly developed, but Primitive Toranian memory remained strongly in place. Much faith in old memories; more of a faith in, rather than power over, life force.

Many will probably argue that we cannot know much about the memory of ancient humans or animals. But, conversely, it is reasonable to infer that humans only began to create written records of the past when their memories became conscious, and when oral traditions went into decline. 'As late as the fifth century ... Socrates laments the spread of reading and writing on the grounds that it weakens man's power of memory'.[29] (Today we lament that the telephone and computer undermine the ability to write.) How similar this sentiment is, in principle, to Steiner's claim that while humans had 'clairvoyant perception of the past, *the need for written records did not exist*'.[30] (original italics) Despite the seemingly bizarre 'symbolic' cultural names used, Steiner's observations are nevertheless consistent with a careful reading of history, and indeed some of the cultural terms are still in use in various contexts. For example, around AD 80, the Jewish historian Josephus, wrote proudly of his culture's written records dating back to the Pentateuch (around 900 BC) which, in his opinion, clearly showed, that they had developed written history long before the Greeks.

Such ruminations have deep implications which lead us to consider the origin of language (see Chapter 9) and the tradition of regarding the 'word' as sacred. The reasons are perhaps simple in as much as they give us a glimpse of emerging self-consciousness. From the beginning language was used to establish a relationship with objects in the sense perceptible world, and so, it follows, to literally 'name the world' and develop a relationship with it: that self-other relationship discussed in Chapter 1. What concerns us here is how memory also developed and helped with the emergence of time consciousness, or the so-called 'Discovery of Time'.

So Toulmin and Goodfield agree that the Greeks could not grasp the historical timescale of creation and so, for them, the creation was still something 'flat'.[31] (This interesting use of a spatial metaphor for time is typical of modern usage, as when we say a destination is two hours away.) Likewise, according to William Manchester: 'In the medieval mind there was no awareness of time ... In all of Christendom there was no such thing as a watch ...'[32] Consciousness still focussed on the relationship of the individual to the external cosmos which was generally assumed to be eternal and divinely created. Time manifested only in the slow rhythms of day and night, and yearly, seasonal and possibly astronomical cycles. In most of recorded history, prior to the Renaissance, only a few scholars had written about the deeds, and accomplishments of great men and even fewer women. Many of the protagonists in these mostly religious or supernatural chronicles dealt not with mere mortals, but with gods, prophets and other divinely appointed emissaries on earth. Most people were illiterate and few cared too much about calendars, which were, as we have seen, based on many different, and complex, confusing chronologies. In short, the development of what we consider to be accurate and precise historical time scales is a very modern invention. The same applies to accurate surveys of land, sea and the heavens. Likewise, accurate measurements of prehistoric, geological and astronomical time scales, and the ancient spatial landscapes of the Earth, solar system and universe are even more recent innovations, still undergoing constant refinement.

The ability to measure time and space with the type of accuracy we take for granted today, has proceeded on a broad front. On the short end of the scale, awareness of the utility of such abstract concepts as hours, minutes and seconds, has only come into consciousness in the last few centuries, as these time divisions proved useful tools in modern science. For example, one cannot calculate longitude — a spatial measure — without accurate timekeeping. Similarly the study of everything from physics and chemistry to biological systems and sports usually requires the simultaneous and accurate measurement of both space and time. Imagine a moon shot where the timing of the launch and the precise spatial trajectory were not coordinated, or an Olympic race where the timing and distance were not both precisely measured. On the long end of the scale, the measurement of deep geological and astronomical time is essentially a twentieth century invention. In much the same way as our ancestors took their first steps into the 'historical' dimension of time, as orally-transmitted memory penetrated back for a dozen or more generations, so geologists and astronomers are now penetrating billions of years of deep 'pre-historic' time. Obviously the time scale is different, but the experiential connection is also qualitatively distinct. Oral traditions begin with the actual experience of direct contact with parents, grandparents and great-grandparents, and the memories of previous generations passed down from 'direct' memory — albeit subjectively filtered to various degrees. However, in the case of deep time, Jurassic Park is resurrected by a complex process of reconstructing fragmentary physical remains of creatures and landscapes never directly experienced by the investigator. It is therefore an interesting paradox that modern science often treats oral traditions as unreliable subjective mythology, whereas the same science has considerable faith in the indirect methods of analysis which re-create remote worlds far removed from direct experience.

Regardless of the problems inherent in evaluating our own subjective and objective analyses, one thing is clear: direct and indirect memory has come a long way. A few millennia ago, humans began their journeys in time as individuals and small family groups bringing the immediate family tree into the fold of consciousness. These oral traditions blended in the deeper past with so-called 'creation stories'. Today, modern science has created an indirect collective memory of the genetic, geological and astronomical past, that gives us the latest 'scientific' iterations of the creation story. But this is, in our opinion, is only the product of a particular and transient paradigm of scientific-rational consciousness, and many scientists admit it is a form of story-telling.

In a very real sense our changing consciousness somehow created the world as we know it by evolving from an unconscious 'without memory' state to a conscious 'with memory' structure. That is to say we fell into the dimension of time. To elaborate on Thompson's observation, noted above, we are dealing with 'the Fall of the soul into time, the entrapment of an angelic soul in the body of an *Australopithecus afarensis*, or the fall of an unconditioned consciousness beyond subject and object into the syntax of thought'.[33] Here Thompson hints at a convergence of events: the fall into time, and the entrance of a new, incipiently individuating, self-consciousness (and time consciousness) into the stream of unconscious nature. This is very close to the sentiments of Owen Barfield when he stated that: 'The dawn of history represents the incursion of a consciously directed human process into the stream of an unconscious natural one'.[34] So, as far as we know no one set out to 'discover' time. Rather, time-consciousness is an emergent property. As Sir Arthur Eddington put it, somehow time, unlike space which we recognize through our external senses, takes a direct 'shortcut into our consciousness'.[35]

As western culture became conditioned to the idea of linear time as a convenient way to organize history books and present geological timescales, various authors began to play imaginatively with the subject. Classic, science fiction stories like *The Lost World* and *20,000 Leagues under the Sea* began to emerge and take us on exhilarating and imaginative journeys through time and space.[36] In the same vein *Gulliver's Travels* and Edwin Abbott's *Flatland: a romance of many dimensions* explored new dimensions of space, while in *Alice in Wonderland* Lewis Carroll cleverly distorted both space and time with what seems like a whimsical, childish imagination.[37] But behind the fantasy a serious scientific consciousness was at work: Carroll's real name was Charles Dodgson (1832–98), an Oxford professor of mathematics. Carroll, like Henri Poincaré (1854–1912) and other eminent mathematicians of the day were interested in the new science of topology, sometimes described as 'rubber sheet geometry' in which, counter-intuitively, many spatial properties of shapes and lines, remained unchanged regardless of the deformations applied to them by folding and distorting the sheet.[38]

All this led the way to Einstein's revolutionary theory of relativity in which space-time is ultimately seen as a continuum. Space is not a three-dimensional phenomenon separated from a fourth, single linear dimension of time. Space and time 'arose' simultaneously, so to speak, as the Big Bang created the expanding universe. So, difficult as it might be for the non-physicist to comprehend, space, time and matter do not have separate existences as Isaac Newton originally supposed.[39] As objects accelerate 'through space' towards the speed of light they increase in mass and their experience is that time slows down. So, in a famous thought experiment, someone could travel for seventy years, at near light speed, visiting a distant planet and returning to Earth to find that they had hardly aged at all while the Earth residents were seventy years older. Time, therefore, is relative to human experience, and not an independent property.

It is hard to get away from the idea that time 'goes on' and has a 'direction' from past to present and future: what has been called time's arrow.[40] The fact that our consciousness recognizes this duration (*dureé* in the philosophy of Henri Bergson) '*wherein we see ourselves acting*' is hard to ignore.[41] For the purposes of understanding how our perception of dimensions can change, we take as our final example Edwin Abbot's story of Flatland where all inhabitants are two-dimensional shapes like triangles, squares and circles, living in a plane. They can only perceive each other's shapes by feeling one another, not by looking from outside. One day a lucky inhabitant escapes into space ('space land') and realizes he is a three-dimensional sphere, and only 'looks like' a circle when he passes through the plane of flatland. Here very simply are two different dimensional perspectives on the same phenomenon. After reading *Flatland,* the well known British author J.B. Priestley extended the analogy to ruminations on time.[42] Priestley had also read *An Experiment with Time* by J.W. Dunne, and was intrigued by the serialization theory.[43] He wondered therefore if one could not use the changing pattern of circles produced by a moving sphere passing through flatland as a metaphor for how we apprehend the fourth dimension of time. Like a glow worm moving over the surface of a large statue we never see things as they are. Instead we only see a series of 3D cross-sections analogous to slices of time, past, present and future.

In summary, we have examined both individual and collective perceptions of space, time and consciousness. In Chapter 3 we focussed on aspects of space-time consciousness as they affect our individual ontogeny. Growth involves a spatial development of physical anatomy (or form) through time that is coincident with the more subtle and intangible shifts of consciousness, which are described by psychologists like Piaget as emotional, cognitive and moral development. We shall have more to say about this in Chapter 6 where we try to integrate our understanding of mind-body (or consciousness and the body) more deeply. But before we do this we need to further investigate how changing structures of consciousness through time may help us understand the evolution of consciousness throughout our species history.

We hope that we have established beyond reasonable doubt that individual memory has evolved through time. Many credible researchers have claimed that our animal predecessors, and ancient hominid ancestors probably lacked conscious memory, and that they lacked a modern sense of time. Slowly, through oral tradition they developed a historical memory of small familial and tribal groups. With time this memory widened to include a written historical record that took in events of regional and eventually global significance. But here we are talking only of what we refer to as conscious memory. What of the memory of Jung's collective unconscious — what Steiner calls the 'group soul' — and what in general parlance we rather vaguely call 'folk memory'?[44] Again there can be no doubt that evolutionary science recognizes profound genetic, morphological, physiological and behavioural connections that pervade the hominid, mammal, vertebrate animal and biospheric web of life. Note here that biologists like Rupert Sheldrake have pointed out that growth is a type of recurrent behaviour, or habit, because each new generation produces offspring, similar in morphology and behaviour, to the individuals of previous generations.[45] Indeed, anatomical similarity is the basis of all our biological and evolutionary classifications.

Again there is no doubt that complex and as yet poorly understood organizational principles somehow remember to keep biospheric physiological systems in extraordinary and intelligent evolutionary balance, harmony and homeostatis from generation to generation and from one geological epoch to the next.[46] So the organizational principles or dynamics 'remember' to operate in such a way as to conserve all the basic elements of biological and behavioural structure of any species, while allowing subtle novelty to be introduced (as Fawcett suggested: see Chapter 2). The very act of reconstructing the evolutionary web of past life through our evolutionary and anthropological science is in itself a collective intellectual act of memory, which uses present 'conscious' consciousness to excavate Jung's collective unconscious. Animals may not have ever been conscious of their past, evolutionary history, despite being deeply anchored to it on a subconscious biological level. But *we are now conscious of our past*, as our culture shows. Even the disparate views of creationists and evolutionists about how to 'explain' our origins, are in essential agreement over the fact that we had past origins!

In our opinion, the metaphor of post-religious spirituality points to the next phase or potential of collective human evolution. Our ontogeny and our species phylogeny strongly suggest that we are a complex mix of physical, emotional, cognitive (mental) and spiritual bodies. All three of the latter components are essentially far more intangible, or psychological and energetic, than they are physical or material. Bearing in mind that the physicists have already told us, for more than a century, that matter and the physical world are essentially an illusion, and that we would be better advised to view the world/ cosmos as a web or energy and complex fluctuating fields, we hope to be forgiven if we suggest that the study of consciousness and psychology, is likely to have to embark, as it already has in some quarters, into the field of subtle energies and parapsychology if it is to make progress in understanding memory and the collective unconscious.

This is not the place to embark on a discussion of past life memories or other para-psychological phenomena, which we can touch on later, once we have completed a discussion of more empirical biological and cultural evidence. However, we must stress that there is really no reason to fall into the dualistic trap of dividing knowledge into scientific and non-scientific categories. The distinguished biologist E.O. Wilson suggested in his provocative, but highly debatable book *Consilience*, that in future there would be a battle for hearts and minds between empiricists (that is, predominantly down-to-earth or materialist scientists) and transcendentalists (those ostensibly more prone to unconventional, mystical or 'way out' ideas).[47] Wilson predicted that the empiricists would inevitably win — a point of view branded by critics as scientific magisterialism, 'scientism', or less charitably as scientific chauvinism.[48] Wilson has the right motives (bringing disciplines together) but the wrong methodology (combining disciplines under a single set of rules). Suppression of pluralism does not work in families, for example, where we have different rules for the perceptions and consciousness structures of children and adults. There is very little reason to support his Scientific Magisterial view and even less philosophical justification. Even in the short history of modern, western science, since the Renaissance and the Enlightenment, there has been an extraordinary collective effort to understand the human body-mind, soul and spirit, by approaching self-analysis from every possible angle (physical, chemical, psychological, artistic, cultural, sociological, historical, anthropological, and so on and so forth). We would rather hope (see Chapters 8–9) that the future holds the promise of a blending of hearts and minds within integral or synergistic consciousness structures that embrace a transcendental empiricism: or is that an empirical transcendentalism?

5. The Evolution of Human Consciousness from Deep Time

A fleeting glimpse of Homo habilis

In Chapter 4 we explored the relatively recent history of our species, and some of the ways various scholars have characterized a cycle of emotional and intellectual moods, mostly measured, somewhat qualitatively, on the scale of generations and centuries. Despite giving us significant insight into how our ancestors may have perceived space and time, few of these contributors said much about deep archaeological time. Richard Keck used the term Paleolithic (meaning old Stone Age) which he characterized as Epoch I — the time of the Mother Goddess — but he did not tell us what it was really like to meet *Homo habilis*. In this chapter we shall be more explicit about studies addressing the evolution of consciousness and culture throughout human history and prehistory. These studies even use linear time scales to categorize successive cultures and consciousness structures. However, as we shall see, linear time scales have certain limitations which some scholars seek to avoid or treat with great caution.

So in this chapter we do our best to avoid the pitfalls of laying out an artificial and strictly ordered chronology of consciousness structures that take us back to the dawn of human pre-history. Many consciousness structures exist simultaneously today, and presumably did in the past. Rather our emphasis is on how to interpret the consciousness structure of individuals or groups from ancient cultures that probably differed fundamentally from to our own. The ideas of intelligent thinkers on this subject incline to the view that there is not a 'shred' of evidence for projecting our thinking into the ancestral world. In fact some scholars infer that in some ancient cultures the concept of thinking did not even exist, or at best was radically different from our own. Admittedly it is difficult to imagine, or 'think' about, another culture without using our accustomed thought processes. But, paradoxically, in the simple act of recognizing that another approach may be necessary we open our minds to other possibilities. Taken to its logical conclusion, we must allow that it is more likely that our present standards, values and thinking are wrong, than that they are flexible enough to always interpret accurately the evidence revealed by past cultures very different from ours. We must try to empathize with, cultivate or experience 'other' consciousness in the way we might when joining in with a child's play or a cultural milieu quite different from our own.

This way of thinking reminds us of Einstein's warning which we adopted in Chapter 2 as a useful mantra: 'We cannot solve problems by using the same thinking we used when we created them'.[1] In a similar vein, Owen Barfield pointed out that: 'We tend to forget ... [that] ideas have changed because human consciousness itself — the elementary human experience about which the ideas are being formed — ... has itself been in process of change'.[2] Two archaeological examples, relating to two different hominid species underscore just how much our scientific consciousness can evolve. First, consider the large Acheulian stone 'hand axes' attributed to the pre-*sapiens* species *Homo erectus*. These were the first ever made with tear drop-shaped outlines, similar in proportion to much smaller and much later spear and arrow heads. Many appear far too large to serve as functional weapons, as old theories assumed. For example, the Furze Platt specimen, found in England in 1919, weighs almost 3.5 kilos (7.5 pounds) and is almost 40 cm long (Figure 5.1). Such large artifacts, and there are many, especially in Africa, may have had some quite different ritual, aesthetic or symbolic meaning. It is interesting that such finds have helped to stimulate the new field of 'cognitive archaeology'. John Feliks, who is also deeply interested in how hand axes appear consistently proportioned to show the 'golden ratio', a mathematically intriguing ratio known as *phi* (φ) (see again Figure 5.1), writes that 'twenty years ago the idea of "cognitive archaeology" ... seemed an impossible pursuit ... [but now] ...

researchers are coming into the field in droves from every other field imaginable'.[3] It seems rather obvious that such excitement over our ancestors' consciousness and potential cognitive ability is a reflection of a paradigm shift in our own consciousness! It is also further evidence of our perennial interest in our cognitive and psychological heritage. However, when all manner of natural objects from sea shells to our finger bones show the *phi* (φ) ratio, the question becomes: do almost one-million-year-old hand axes indicate a conscious effort by *Homo erectus* to create mathematical proportion, or was the design created on a more unconscious or instinctive level?

Likewise, our second, more recent example considers how archeologists have changed or doubted their previous interpretations of 10,000–30,000 year old Late Paleolithic cave sites and cave paintings attributed to our own species *Homo sapiens*.[4] Instead of regarding them as simple shelters where artists could depict hunting activity, many now prefer to infer that such sites were often far deeper than needed for shelter, and may therefore have been used to perform shamanistic rituals. It is contemporary anthropological interest in shamanism as much as any startling new physical evidence that encourages us to project shamanistic interpretations on to our cave-dwelling ancestors. Of course, many interpretations are possible and none are certain. As the anthropological literature proves, speculation is rife, but at the same time very intriguing.

Let us take the opportunity to test our ability to understand prehistoric consciousness and culture, and perhaps even to face the challenge of pondering the consciousness of Neanderthals and other species such as *Homo erectus* and *Homo habilis* that predated the emergence of our own species *Homo sapiens* some 100,000 years ago. This is perhaps not so difficult, or different in principle, from the training demanded by any discipline requiring a specialized focus.

The scene is about 3.5 million years before the present (BP), the place the east African savannah. A group of three small upright apes leave clear footprints in a blanket of damp volcanic ash as they walk northward. Two individuals are taller than the third, but even these two are less than four feet tall. One individual holds a nondescript

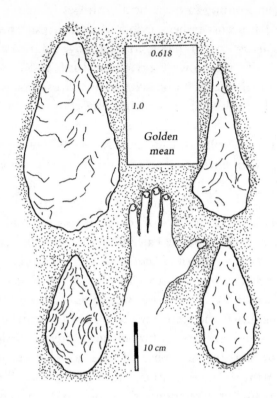

Figure 5.1. The large size of million-year-old Acheulian hand axes from Europe (top) and Africa (below) are attributed to Homo erectus. Many appear too large to be used effectively as hand tools or weapons. Large tool (top left) weighs 7.5 pounds. See text for details.

rock, another an equally nondescript stick. The group could possibly belong to the species modern anthropologists call *Homo habilis*, but it is difficult to tell at a distance. Perhaps they belong to the species known as *Australopithecus africanus*, or perhaps *Australopithecus afarensis*, a closely related species whose best-known fossilized representative was affectionately christened 'Lucy' by twentieth century anthropologists. What do we know about these creatures we claim as our ancestors? What did they think, feel or have to say for themselves? Did they even have a concept of thinking, feeling, speaking or 'selves'?

Most anthropologists and readers of anthropology find it easy to imagine this scene. There is a certain evocative nostalgia in creating images of lost ancestors. To use a popular TV documentary metaphor, we can 'walk with' our distant relatives — even literally walk in their footsteps. All this mind's-eye imagining across the gulf of several million years evidently shows that we have a great curiosity and an affectionate affinity for these extinct creatures that ostensibly link us to our origins. Why? Is it merely because they share one of our most human physical characteristics — upright posture? Or is it simply that insatiable intellectual and spiritual curiosity about our origins impels us to create such images?

How accurate or realistic are such images, reconstructed from only a few bones and stones, without the benefit of direct contact or the 'fantastic' luxury of time travel? We actually have no complete skeletons, or evidence of skin colour, degree of hairiness or nakedness. Do we cherish our reconstructions because they somehow resonate with our deep genetic memories? How different is the evocative image, now deeply imprinted in our minds, from the images that a psychic archaeologist might 'see directly' or 'experientially', though perhaps somewhat darkly, through a glass clouded by the mists of time?

Let us reformulate the last three questions so as to shed a clearer light on the cognitive methods available for accessing the past:

1. One can reconstruct past scenes from limited physical evidence and clothe them with the 'thought-experiment' imagination characteristic of our present consciousness. This involves creative, mental time travel, and considerable confidence in our reconstructions.
2. One may, in rare cases, have direct psychic (experiential) insight into the past, and claim the experience as an authentic case of psychic time travel.
3. One may intuitively judge methods 1 or 2 as right or wrong, and accept or reject them with or without strong justifications. How any intuitive 'sense' of genetic memory could be validated is an absolute mystery.

All three methods have something in common. They show our psychological predisposition to explore the past and create explanatory interpretations. But they are also all open to error, criticism and re-evaluation. The first thought experiment method allows the luxury of imagining scenes for which much evidence is lacking. Even if time travel were in fact possible, allowing a meeting between anthropologists and three million year old ancestors, what would be the result? Even with a hypothetical common language, or other means of communication, it is doubtful that either group could learn much from the other, without actually communing with the other's consciousness. To tell a three million year old creature it was a species of *Australopithecus* or 'ancestor' (from the anthropologist's past), would be not just meaningless, but wrong. Even if the species had named itself, which is only a hypothetical possibility *if* they were self-conscious, their chosen name would certainly not be *Australopithecus*. Anthropological concepts of past and present would have already been rendered meaningless as a result of the purely hypothetical time travel experiment. Thus, the anthropologists themselves would be confused, and unable to use their linear time vocabulary. Likewise, to ask *Australopithecus* whether their sticks and stones were tools, or religious symbols would be equally futile unless they shared our concepts and technological and religious vocabulary. They 'might' understandably regard us as apparitions from another world. And indeed the hypothetical time-travelling anthropologist would be exactly that — an apparition, fantasy or, by our own standards a 'scientific fiction'. Ironically, the only potentially plausible direct 'experiential' communication between a modern human and an ancestor is in the intangible realm of the psyche. Hard as it might be for sceptics to accept, there are authenticated cases of clairvoyant or psychic detectives like Gerard Croiset (1909–80) who solved dozens of 'hot' and 'cold' criminal cases by reaching through time to 'see' past events. Understandably such authentic psychic powers, although rare, attract the attention of police, psychologists, parapsychologists and sceptics alike.[5]

The Gebserian paradigm: an epic cultural philosophy

As a philosophical, Polish-born German of high breeding, Hans Gebser (1905–1973) was a quintessential European intellectual with an extraordinary depth of scholarship and a profound appreciation for cultural diversity. As a young adult he changed his name to Jean (the French equivalent of Hans) and later moved to Italy, then to Spain where he worked for the Ministry of Education and became friends with the celebrated and flamboyant poet-playwright Federico García Lorca who was killed by Francisco Franco's fascist regime in 1936 at the outbreak of the Spanish Civil War. Gebser was also nearly killed by a bomb twelve hours before he fled to France, where he pursued his literary studies and his ideas on how changes in language were heralding a new 'aperspectival' mode of thought. Then, in 1939, two hours before the border closed, he went to Switzerland where he immersed himself in his cultural studies. There among others he befriended the great psychologist Carl Jung. Deeply interested in the European cultural differences between 'intellectual' Germans and 'emotional' Spaniards, Gebser extended such comparative cultural thinking to an exploration of Asian culture and in 1949 produced his great work *Ursprung und Gegenwart,* on the evolution of consciousness. This book was translated into English almost forty years later under the title *The Ever-Present Origin.*[6] Even in translation this title evokes the essential enigma of consciousness which somehow arises or emerges spontaneously like an inexplicable organic force at the core of existence. In German *Ursprung* has more of the meaning of 'spring' or 'eruption' than 'start' or 'beginning'.

For Gebser human consciousness was: 'Neither knowledge or conscience but ... in the broadest sense wakeful presence'.[7] He described five consciousness structures — Archaic, Magic, Mythic, Mental and Integral — each with their own distinctive manifestations in the history of human culture (see Table 5.1). He believed that the structures were not discrete sequential stages in time but a more cumulative unfolding or intensification process 'towards enrichment of structure' — a type of complexification or 'plus mutation', creating 'dimensional increment'.

By contrast, Gebser saw the concept of Darwinian evolution as a rather negative 'minus mutation' making organisms overspecialized, passive victims of environmental pressure.[8] So, he saw consciousness evolving in a synergistic mode that differed from the development of biologic systems. In fact, he disliked the concepts of development and progress and saw that seedling ideas often appeared early, then disappeared only to blossom later. As Table 5.1 shows, Gebser was fairly specific about the relationship of emergent consciousness structures to organs of the body and to the modes of awareness that we call sleeping, dreaming and waking. As modern scientific vocabulary might put it, new order and complexity emerge from chaos.

In Gebser's view, new consciousness structures or mutations emerge from the ever-present origin or source rather abruptly as in punctuated equilibrium (or spiritual epiphany). Old structures are displaced or dethroned but remain in more or less latent form in each one of us. So the past is integrated with the present. Thus, Gebser's work is not merely retrospective, but also introspective.

Table 5.1. Gebser's consciousness structures and their relationship to selected biological, physiological, objective, subjective and cultural parameters. (Modified after Gebser 1986.)

Consciousness structure	Organ emphasis	Consciousness state	Ojective/ subjective focus	Sign or Symbol
ARCHAIC	—	Deep sleep	Unconscious spirit / none	none
MAGIC	Viscera — Ear	Sleep	Nature / emotion	point
MYTHICAL	Heart— Mouth	Dream	Soul or psyche / imagination	circle
MENTAL	Brain — Eye	Wakefulness	Space world / abstraction	triangle
INTEGRAL	(Vertex)	(Transparency)	Conscious spirit / concretion	sphere

Gebser draws on a wide range of cultural and linguistic evidence to show this organic evolution of consciousness proceeding inexorably throughout history towards the present emphasis on mythical and mental structures that prepares the ground for the integral structure. His masterful reading of the text of the world *(logos)* created by humanity is another archeological excavation of consciousness. Gebser retraces the emergence of all the esoteric symbols (art, language, technology) created throughout history. For Gebser, 'integral reality is the world's transparency, a perceiving of the world as truth: a mutual perceiving and imparting of (the) truth of the world and of man and of all that transluces both'.[9]

To approach this integral and concrete appreciation of reality, we must understand our individual and collective experience of these various consciousness structures in the context of our own existence and the history of humanity. The first step is to familiarize ourselves with different consciousness structures that have their own psychological traits which are perhaps as distinct in meaning as the syntax and vocabulary of different languages.

Archaic consciousness: the dormant soul

In modern English, the prefix 'arche-' usually means 'old', as in archeology or archaic, but it can also mean original or first, as in archetype. The Greek root *arkhe-* has the connotation of inception or original structure.[10] Gebser stresses 'origin' as ever-present, rather than inception which implies a beginning point in time. For Gebser 'origin' implied an eruptive emergence, like the source of a river, or the origin of species — an eruption of novelty. The origin of life or species is a mysterious process by which a new human life comes into existence as a constant manifestation of the ongoing creation of the species. (At least two human babies are born every second.) Gebser cites Chuang Tzu as saying that: 'Dreamlessly the true man of earlier time slept'.[11] Thus, the soul was dormant. The archaic consciousness structure is a type of 'unconsciousness' or simple consciousness similar to that of the infant before self-consciousness emerges, and perhaps similar to the unconscious but 'intelligent' instinct of plants and animals.

Such consciousness structure is timeless and so it is meaningless to use temporal language in reference to the archaic structure. Early humans and ancestral 'animal' species evidently did not perceive the passage of time. As with infants, their existence was fully and unconsciously integrated with the intrinsic, organic rhythms of nature — an undifferentiated part of the psychic web of the biosphere: that is, of universal consciousness.

Before humans had created any distinctive cultural artifacts such as tools or art, they were psychically undifferentiated from nature. In such a state they had no separate identity. Yet they functioned or interacted instinctively with the environment like animals or sleepwalkers. One might well ask how one can make such bold assertions about 'another' structure of consciousness. One compelling line of evidence is the lack of cultural artifacts that express a separate identity. Like the infant the soul was dormant in the archaic individual — asleep and yet to awaken. Individual identity was deeply ensouled in nature. Inner and outer worlds were a completely undifferentiated 'macrocosmic harmony'. As we saw in Chapter 4, only later did the emergence of the perception of an individual human soul impact on human history.

According to Plato: 'The soul ... (came into being) ... simultaneously with the sky', in other words when people made a distinction between heaven (sky) and earth (plant life).[12] In early Chinese chromatic symbolism, the word *ching* is used for the colour of the sky as well as for the sprouting plant. Plato's idea that Ideal forms (of thought or mind) came down from the higher realm, heralds a long history of heaven-earth dualism in western thought.

Religious studies expert Mircea Eliade notes that heaven and sky are intimately associated in many cultural and linguistic traditions including our own.[13] The fact that we now distinguish heaven as a spiritual symbol distinct from high atmospheric 'space', such as 'sky' or stratosphere, only serves to stress that our consciousness is different from the original human experience before humankind 'fell' to earth through the experience of self-knowledge. Nevertheless, words like 'high' and 'higher' retain distinctly positive, uplifting connotations and are often religious or spiritual rather than merely spatial in their implications, as in high ideals, higher self or higher consciousness.

So, archaic consciousness is zero-dimensional according to Gebser. We must conceive its structure without reference to our experience of space and time. This also means that material and immaterial, tangible and intangible, concrete and abstract are not differentiated. Archaic consciousness may converge with the Buddhist concept of *mu*, or emptiness. This does not imply that Buddhism is a manifestation of archaic consciousness. Rather, Buddhism is a practice that actively explores many consciousness structures and can more fully appreciate their diversity. Our typical experience is not the only mode of consciousness. Indeed it is precisely this rigid 'one-consciousness-fits-all' perspective that we want to avoid. We are all too familiar with the pitfalls of this type of cultural myopia. It is unrealistic to expect children to have an adult view of the world, and hopefully having once experienced it themselves, adults can appreciate infant consciousness and its 'archaic' characteristics. So rather than use adult 'reason' we evoke feelings and images. This same process applies to understanding archaic, magical and mythical consciousness structures which cannot be understood or appreciated only with the language of thinking and reason.

Magical consciousness: the emergent awareness of nature

Several consciousness mutations took place before thinking or mental consciousness evolved from its archaic, magical and mythical precursors. These mutations involved the emergence of will, feeling, thinking and intuition. These processes were just as complex and unconscious or subconscious as any we go through as individuals. Collectively these mutations have helped shape the evolution of human culture and our orientation in space-time. But as Gebser warns, we should guard against 'relativizing' the process to suggest that the changes were directed towards us. While this fear of 'goal oriented' or teleological thinking is common among scholars, it is, as we shall see, a very common mode of thought.[14]

In part, the magical consciousness structure represents the emergence of the will. The word 'magic' derives from the ancient root *magh-* (to have power), related to the German *macht* and to the English 'make' and 'machine'.[15] These terms are strongly connected with the emergence of physical power in the human. The child or individual is 'wilful' when it is determined to exercise its power on an object in the world. It is this focus on one thing or one point, selected from a myriad of worldly possibilities that gives the magical consciousness structure what Gebser calls its one-dimensional characteristics. Thus, strong-willed individuals have climbed Everest, won Olympic medals, lifted cars in moments of crisis, or, in the context of early cultures, felled mammoth with wood and stone weapons. But this 'will power' does not come purely from conscious internal decisions.

The anthropological literature is full of examples of what is called sympathetic magic — popularly known as voodoo in some cultures. Leo Frobenius gives the example of African pygmies who draw an antelope in the sand that is struck by a sun's ray in a certain spot on the neck.[16] They then shoot an arrow into the same spot and later during the hunt kill the animal with a wound to the very same point. Gebser calls this magical structure 'punctiformal'. (See Figure 5.2.) Other forms of magic are still practised today with evil consequences. For example, in 1992 a New Guinea newspaper reported that two individuals quite overtly admitted killing the spirit of a third party without laying a hand on him.[17] Not only did the individual die quickly, but the confessed perpetrators were given one year jail sentences under the Sorcery Act. Such practices surprise and confuse our mental-rational consciousness. Fear or ignorance of the psychological dimensions of voodoo or shamanism lead to doubt or scepticism, although the frequent use of the subject in movies, suggests that considerable interest in the magical consciousness structure still persists in the age of electronic media.

Paleolithic cave paintings, dating from about 10,000–30,000 BP, have also been interpreted as evidence of both sympathetic magic and shamanism (see Figure 5.1). Shamanism is again mysterious to the modern mind, but it has been intensively studied by anthropologists and is widely regarded as the archetypal or 'original' religious or spiritual practice. The shaman is someone who can travel into the unconscious or spirit world, often 'experientially' adopting the identity of an animal in the unconscious domain and symbolically in the conscious world.[18] The shaman does this by going into a trance or

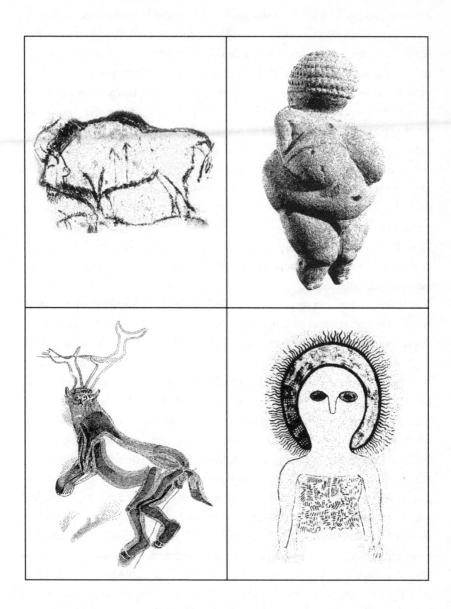

Figure 5.2. Representative magic and shamanistic cave art. Top left: bison and spears. Top right: typical Venus figurine. Bottom left: the famous 'shaman' depiction from Trois Frères cave, France. Bottom right: a figure with a large head and no mouth (after Gebser 1986, Fig. 9): compare with Figure 3.2 above.

ecstatic state. (Ecstasy literally means to 'stand outside' oneself.) This transformation of consciousness allows direct experience of 'another' state. Like the medium, whose body is temporarily occupied by the spirit of another, the shaman is indeed in an altered state, with the physical symptoms of trance and unusual body motions to prove it. We will have more to say about this below, but for the meantime, we should not forget that we are capable of projecting our imaginative thoughts to other places and people, even if most of us do not 'take on' or inhabit the consciousness of another entity. (Nevertheless, it is quite a nifty psychic trick to be able to project our thoughts elsewhere. Archaic and magical consciousness structures probably did not share this same cognitive faculty.)

Whether cave paintings are examples of shamanistic or sympathetic magic, and whether practitioners of such magical rites could exercise power over animals is for the reader and anthropologists to debate. The connection between hunter and animal, individual and object, was likely not viewed as 'causal' in the magical consciousness structure. As human awareness of the outer world dawned, humans were at first highly sensitized to it. All external forces, indeed the entire world in which humans were embedded began, for the first time to have powerful symbolic meaning. Gebser suggests that the injection of Ego consciousness into the unconsciously bound vital energies of the biosphere breaks or 'disrupts' the spell of nature and the incipient urge for freedom pits man against nature. The beat of the drum provides an example of the means to challenge and dominate nature by drowning out the sounds of the jungle. In a sense our archaic and magical ancestors were, like very young children, awed, even frightened by their new sensory awareness of an unfamiliar world, and still immersed in a sleeplike state — in a complex nexus of subconscious and unconscious forces and without awareness of their potential for motivation. Humanity was 'coming to its senses' for the first time, trying to make sense of a complex external world which the human vessel was, for the first time, beginning to consciously and internally digest.

Though alien to our present consciousness, in retrospect this perception makes perfect 'sense'. How could it be otherwise? If humanity was perfectly integrated with nature in macrocosmic harmony there was no 'other' world out there to experience.[19] If you do not have self-awareness, you cannot associate sensory experience with the self. By definition you have no experience of your physical body, much less emotions or thought. This thesis is developed by Barry Long, who has much to say about the psychological birth trauma involved in the 'fall' from what he calls the unconscious 'psychic web' of the biosphere into self-consciousness.[20] In a parallel sentiment Gebser shrewdly points out that in the pre-self-conscious state, individual power was not felt as an internal force of will freely exercised. Rather 'in the hunting rite, the egolessness is expressed first of all in the fact [that] the responsibility for the murder, committed by the group Ego, against a part of nature, is attributed to a power already felt to be standing outside: the sun'.[21] Again, in retrospect this makes good sense. No individual can take responsibility if no sense of individuality exists. The operative forces are external and truly felt to be in nature. The human merely participates in the macrocosmic force field, and the sun, for example, animates all nature from flower petals to the dawn chorus.

We see these dynamics play out every day with children who explain accidents as the result of supernatural forces. The glass just jumped off the table and broke. My brother, or even an imaginary friend, made me do it. Not only do young children readily attribute events to magical 'causes' and suffer quite genuinely from the loss of cherished symbols and objects, but they are proven quite capable of relieving pain through magic rituals. Kisses and band aids heal pain magically. As adults we suggest the 'placebo' effect thus defining a cause that is quite irrational and purely psychological.[22]

Adults are just as susceptible as children to invoking unknown forces. Our language is riddled with excuses. 'It fell out of my hand,' or, 'It got away from me,' instead of, 'I dropped it'. The stockbroker blames 'market forces' for the loss of your investment. We are torn between self-conscious responsibility and our tendency to invoke magical, natural forces (gods and supernatural powers), for which we cannot feel responsible. Our magical consciousness structure is alive and well.

Gebser considered the magical consciousness structure to be pre-linguistic. He noted the lack of a mouth in many early rock carvings and sculpture motifs. One listens to the world rather than speaking to it. Schizophrenics hear voices independent of objective reality. So too do modern mediums, many of whom are highly telepathic.[23] Laurens van der Post described similar telepathic abilities among the San or !Kung 'Bushmen' of the Kalahari.[24] This, again, is an animal-like instinct similar to that of the psychic pet who knows when its owner is coming home.[25] (Ironically, while we tend to admire these hard-to-understand telepathic abilities, we are also inclined to think comparisons with animals and schizophrenics are politically incorrect. Nevertheless, the examples serve to illustrate the intellectual confusion caused by assessing different consciousness structures by a single standard).

When voices are heard there is little or no differentiation between the hearer and the heard. The experience is 'felt' but not rationalized — it is like an oracle or supernatural force.[26] For this reason the shaman and spiritual guru are revered because they receive messages from 'elsewhere' to which the normal person does not have access. All manner of non-vocal or non-verbal communications fall into this category. A mother receives a telepathic message that a child is in danger far away. An individual experiences an extraordinary coincidence, such as waking up at the moment of death of a faraway relative or friend.[27] Our rational consciousness requires explanations or labels, even if they are labels for phenomena such as magic that are neither fully believed nor understood. The use of terms like supernatural and paranormal is highly ironic and revealing. They show that we recognize forces that differ from those we can explain and call normal or natural. It is contradictory to define phenomena while at the same time doubting their reality! By giving them names we give them meaning and bring them into conceptual being, even if our meanings remain mysterious and ambiguous. In the magical consciousness structure they are felt as real but not considered as odd or as separate from everyday experience. By repeatedly trying to define magical consciousness from our contemporary mental perspective, we demonstrate how hard it is to 'get out' of our present consciousness structure.

Gebser identified the magical consciousness structure with the stomach, and to this day we still speak of powerful 'gut feelings'. Some magical practices use entrails to divine the future.[28] 'When Moses wrote of Joseph's "bowels yearning upon his brother", or David prayed to the Lord not to forget his bowels, or when Isaiah, Jeremiah, and other inspired men of old spoke of the "sounding" or the "troubling" of bowels, they all each endorsed the belief prevalent among the Japanese that in the abdomen was enshrined the soul.' Thus, in the Japanese *Bushido* code, the ritual of suicide by evisceration known as *hara-kari* allows the victim to show his entrails to demonstrate that they are clean![29] The viscera have a more prominent and visible position than most internal organs. The stomach is remarkably mobile, especially during pregnancy — also known as 'quickening'. In the so-called fertility cults represented by Paleolithic 'Venus' figurines, the swollen stomach is much emphasized. Thus, the dynamic 'centre' of life and consciousness is identified with the 'lower' organs. This circumstance is quite different from the perception associated with other consciousness structures. The cave painting shown in Figure 5.3 depicts a hunter connected to a mother figure by an umbilical cord — a type of external entrail. It is possible to interpret this as evidence of a strong belief in the magical properties of women in producing new life. The drawing depicts a literal 'gut level' connection associated with the lower, bodily organs of reproduction. As noted below, humankind, at this stage of psychic evolution, may not have known what we collectively call the 'facts of life'. Thus, birth was a magical and awe-inspiring occurrence which conferred great status on the woman.

Mythic consciousness: the emergent time-conscious soul

Our efforts to appreciate and understand the mythic consciousness structure again requires recognizing its inherent and distinctive characteristics and manifestations as an integral part of human consciousness as a 'whole'. Human development and history will again be our guides. The emergent mythic consciousness structure builds on archaic and magical structures in a process of evolutionary complexification rather similar to the building of the mammalian brain and neocortex 'on top of' the reptilian brain stem. So far so good, but we must keep in mind that strict time lines are primarily 'mental' perceptions of reality. As Gebser says 'it is pure speculation if we try to locate something timeless in a temporal framework'.[30] To get

Fig 5.2

Figure 5.3. A Paleolithic rock painting from Algeria shows a female figure connected to a hunter by an umbilical cord. (Modified after Thompson 1981, Fig. 9, p. 122.)

at the essence of the mythic perception, we can explore the childhood development approach. What is the difference between the child's magical world of make-believe — where it is possible to wave a magic wand and make Mommy disappear or turn into a dinosaur — and the next emergent consciousness structure, where imagination focusses on mythical superheroes and larger than life, real life heroes? The emergent mythical consciousness involves individuation and the maturation of the inner world of the soul. As Gebser put it, just as the magic structure brought about 'a gradually increasing awareness of man's individuation ... [and] ... a disengagement from nature and an awareness of the external world ... [so] the mythical structure, in turn leads to the emergent awareness of the internal world of the soul; its symbol is the circle, the age-old symbol for the soul. The individuated point of the magic structure is expanded into an encompassing ring ...' of two dimensions.[31] Here we may remember our discussion from Chapter 4 regarding the cyclic motion of the sun, moon and other planets, as well as that of the night sky and the precession of the equinoxes.

The derivation of the word 'myth' has the dual and ambivalent meaning of mouth, sound or speech (*mythos* in Greek), on the one hand, and, on the other, Greek *myein* or Sanskrit *mukah*, which mean mute. Thus, the mythical structure is intimately associated with the origins or development of language, the 'muse' and self-expression of inner soul forces.

There is a polarity between speaking and being mute. Speech requires both sound and mute pauses, and communication requires speaking and listening. It is a rhythmic process like the cyclic motion that bring us light and dark, night and day, summer and winter. The essence of polarity is differentiation, subject and object, internal and external, within the context of a unified whole — what has ultimately been called polar completeness.[32]

This two-dimensional perception of polarity is evident in much mythic Greek art which shows distinctions between earth and sky. Figures have their lower bodies embedded in nature and entwined in plants while their upper bodies rise skyward separating from nature (see Figure 5.4). Accompanying this experience of separation there is a very real sense of loss of connectedness. Instinctive, psychic, telepathic communication characteristic of the archaic-magical structure dissolves or is suppressed and imagination takes its place. Here is the birthplace of much classical mythology. Humans could represent themselves in the form of gods or spirits that combine human form and human sensory experience with the still powerful forces of the unconscious psyche and external cosmic phenomena like stars, planets and other creatures with which they identified on a deeply experiential level. The result was the 'mythical dream ... so vivid as to be almost identical with what we today call "reality".'[33] According to Gebser it was through the cycle of dream and wakefulness that mythical man came to an awareness of consciousness. Such awareness is captured in Chuang Tzu's famous dream reference: 'Are not you and I perchance caught up in a dream from which we have not yet awakened?' and in many references for the need for the truly conscious man to 'awaken'.[34]

Figure 5.4. Magic figures surrounded by nature and mythic figures half-immersed in nature. (Modified after Gebser 1986, Figs. 5 and 19.)

A recurrent mythic consciousness motif involves heroic voyages of adventure, often over seas, rivers or lakes. The hero archetype, much discussed by Joseph Campbell, is a fairly straightforward symbol of emergent individuality-Ego, jealously guarded against rivals.[35] Water however is symbolic of the unconscious, and the journey across it represents a battle to overcome unconsciousness. When Narcissus sees his reflection he sees his own soul, and by not separating from it he drowns in the unconscious. For mythic man the image (from which we get imagination) is an exterior divine projection of the soul — mirroring its internal counterpart.[36] The Latin for mirror is *speculum*. Hence we have both imagination and speculation: incipient thought processes. Gebser holds that the older forms of the mythic structure are timeless, but as later myths recall 'the genesis of earth and man' so time consciousness emerges.[37] It is perhaps helpful to note explicitly, that our thoughts still mostly arise randomly, incipiently and 'imaginatively' before we consciously grasp and direct them into rational channels. Anyone who has tried meditating will attest to this process.

Owen Barfield's knowledge and interpretation of ancient languages and literature was as profound as Gebser's. As he studied the ancients he came to realize he was studying the evolution of consciousness. He understood that most of the ancients had not separated thought or 'conception' from 'perception', and he understood the difference between the 'inspiration' of the muses and the power of independent 'imagination', which he in turn linked to the polarity between communication and expression. In the same vein he understood that poetry was an emotional response to the natural world, while prose was a rational, abstract and analytical response.[38] Table 5.2 shows us the internal and external polarities inherent in these linguistic and cognitive domains.

Table 5.2. *A synopsis of linguistic and consciousness polarities discussed by Owen Barfield. Note the difference between external and internal qualities that tend to characterize mythic (less conscious) and mental (more conscious) structures.*

EXTERNAL	Poetry	Inspiration	Perception	Communication
INTERNAL	Prose	Imagination	Conception	Expression

Today we associate waking consciousness with the brain, but mythic man, whether a classic Greek or a Native American saw the seat of consciousness in the heart or the breath of inspiration and respiration. Carl Jung famously reported a conversation with a native American Indian from the Taos Pueblo who thought the white man mad, 'uneasy and restless' ... 'always seeking something'. When asked why they were mad, the Indian replied: 'They think with their heads ... We think here, he said, indicating his heart'.[39] With the shift from mythic to mental consciousness or from heart to brain it is possible that the organs of speech were activated, or activated more fully. This is more that mere conjecture about the symbolic rise of consciousness from lower to higher organs. During early human development activation of biological organs follows a similar path (as we shall examine in detail in Chapter 6). The stomach feeds through the umbilical cord and the heart beats long before the first breath at birth and the later activation of organs of speech and reason. The posterior > anterior rise of consciousness is biological as well as symbolic. It all comes together as our theme of integration ultimately demands.

At some point humans must have learned or 'consciously understood' what we call the biological facts of life. There seems to be no getting around that fact that at one time, they did not consciously understand what we call reproductive cycles and 'mechanisms', but that later this knowledge was acquired. In this pre-facts-of-life epoch it makes good sense to infer that women were revered for their magical reproductive powers, as pregnant figurines and umbilical cord depictions indicate. This was the childhood, matriarchal age of the Goddess discussed in Chapter 4, and labelled as Epoch I by Richard Keck. When humans, especially men, learned that they also played a significant role in the reproductive process, masculine power and ownership received a significant boost, hastening the patriarchal age which Keck labels adolescent Epoch II. One can almost envisage the 'macho' exuberance experienced by men as they first began to understand their procreative powers.[40]

Mental consciousness: the measure of man and the universe

The mental-rational consciousness structure is characteristic of much modern western intellectual and scientific thought. Our familiarity with this mode of perception should make it easy to understand. But, if we haven't thought too much about it we may regard it as the *only* meaningful perspective on consciousness. But what may seem like second nature — the way we view the world — is part of the great flux of human consciousness and the even greater wellspring of organic and universal or cosmic consciousness.

From previous discussions we know that archaic, magical and mythical consciousness have zero, one- and two-dimensional characteristics. So, it follows that mental consciousness is three-dimensional. Indeed this mode or structure is primarily 'dimensional'. It sees the world as measurable in every dimension from space and time, to temperature, colour, mass, intelligence, and even consciousness itself. The latter endeavour is, in our opinion, a tall order. Although many traditions, especially those of the East, have studied consciousness experientially, and in depth, Western science is now taking on the task from its own spatial and very three-dimensional 'perspective', looking for the 'location' or 'seat' of consciousness in the brain. Some have not flinched in labelling the human brain 'the most complicated, complex, and wonderful creation in the entire universe'.[41] As Barfield and the Pueblo Indians would remind us, the ancients would never have located thought or consciousness 'inside' the brain.

Art history tells us that perspectival 3D paintings were rarely accomplished before the Renaissance a mere five hundred years ago. Our ability to depict our world in three dimensions and draw parallel lines that recede in the distance to a 'vanishing point' also indicates that the mental-rational consciousness is highly connected to vision and spatial awareness. 'Seeing is believing,' we say and we take for granted that 'perspective' is necessary for an accurate view of the spatial and dimensional characteristics of the world. Gebser argues that between 1250 and 1500 humans were 'possessed by and did not merely possess, the urge to realize space'. He gives the example of Francesco Petrarch's letter describing his ascent of Mount Ventoux in France around 1336 as 'an epochal event [that] signifies no less than the discovery of landscape'. It was a step out of the 'transcendental gilt ground of the Siena masters ... into real space'. In many passages in Petrarch's letter, he speaks poetically of the alternation of his external gaze on the surrounding landscape, which was both exhilarating and daunting, and his inward retreat to contemplate his wondrous soul, transporting him back 'from space to time'.[42]

Gebser notes that the terms mental and rational come from the Greek *meno*, 'to think discursively', and *ratio,* meaning 'to divide' or measure. It is with the Greeks that we see the birth of many aspects of modern western geometric science which rely on measurements of length, time, mass, temperature and so on. It is significant that Plato perceived a higher order of organization in the universe, an ideal mind that was dimly reflected in the imperfect mind of man. Following Socrates, he was smitten by the emergence of this new mental power of mind or reason *(logos)* and its ability to explore or analyze the world or itself.[43] This self-referential ability or ability to self-analyze is one of the hallmarks of philosophy and epistemology. The terms 'mind' and 'man' also have derivations connected with mental capacities and thinking. Since the time of Aristotle humans have appointed themselves as the 'animal rationale'.

Following the mythic awakening of the soul, mental man experienced the further waking of the Ego. Man saw himself as a separate entity in relation to the cosmos. This dualism, as in Plato's distinction between ideal, higher, and worldly, lower forms of mind and thought, fostered monotheism and emphasized the subject-object (man and god) dualism that has so occupied our philosophical thoughts ever since (see Chapter 1). Plato saw the eye as the extension of the head — able to 'grasp' or understand the objects of its perception. The Greeks associated the rational mind with the eye and the brain also. When Zeus has his head split open by the axe of Hephaestus, the clear-minded and knowledgeable Athena is born. Likewise from the seventh and sixth centuries BC onwards, Greek sculptures begin to portray prominent, clean, clear foreheads. It may well be that the pre-frontal cortex underwent significant psychological development at this time, even if the actual anatomical changes were hard to measure in physical terms. Certainly we see no sign here of the sloping Neanderthal forehead and low brow ridges.

The Mental consciousness must discursively think about the objects of visual perception (Chapter 1), and find a way to represent them to the conscious mind. This is done by an analytical system of abstract measures: big-small, short-long, heavy-light, and so on. Anything not consciously seen and available for discursive analysis is an unconscious day dream. Nevertheless, mental man is good at creating abstractions to represent objects as large or small, light or dark. Miles, inches, hours, minutes or shades of infrared red or ultraviolet are not real in any tangible or absolute sense. They are products of our measuring 'ratio'-producing ability to abstract the world through discursive thought (converting reality into mathematics). No child who has yet to attain the age of reason (abstract thought) can bring you a measured litre of milk or a kilo of potatoes.

As Gebser says, the mental world 'is a world of man ... predominately human ... where as Protagoras stated "man is the measure of all things" where man himself thinks and directs his thoughts. The world which he measures ... is a material world of objects outside himself with which he is confronted'.[44] Imaginative abstraction takes over from inspiration. The phase shift in consciousness was experienced by Parmenides when he said 'for the same thing is for thinking and for being'. Thought achieved primacy as an essential facet of being.[45]

Does this imply, one wonders, that in previous pre-mental structures being was primarily a non-cognitive state such as 'feeling'?

The mental-rational consciousness is ascendant in both its strengths and weaknesses. Note the 'measured' dualistic terminology. Gebser himself used similar terms speaking of the 'efficient' mental structure phase and the 'deficient' rational phase. As thought, mentality or reason was first experienced by the early Greeks, as a 'visitation' from the Muse, it gave birth to all the philosophy and modern sciences which we revere as seminal in all our histories of modern western thought. But our way of thinking has often ossified with an over-emphasis on rationalism, analysis, overspecialization and materialism. For too long, modern thought has seen the world as a collection of material objects. Originating with the Sanskrit *ma* and *matra,* we derive the Greek *mater* (Matter), *meter* (Great Mother) and *metra* (measure of external things).[46] The emphasis on matter often favours objective patriarchal culture and suppresses subjective feminine sensibilities. This materialism is intricately related to specialization and intense analysis. Water is two parts hydrogen and one part oxygen, salt is one part sodium and one chlorine. The cosmos is divided into organic and inorganic. Science has separated physics, chemistry and biology from philosophy and literature. Consciousness is perhaps a mere epiphenomenon of the brain, our family inheritance a mere DNA molecule. The soul and psyche are immaterial and so unreal, and imagination is a mere fantasy, a will-o'-the-wisp — not, as Barfield so painstakingly showed, the thought faculty which sets us free. We risk reducing our rationalizations to an absurdity of ever decreasing ratios as we study more and more about less and less.

Here then is an alternate perspective on the shortcomings of the more deficient aspects of our present consciousness structure. This rational analysis of extreme rationality counteracts the hubris of deifying rationality. If we can measure molecules and assign them scientific mass values, we can also assign values to our scientific endeavours. Indeed science is proud of its analytical self-correcting debates. Mental consciousness is undoubtedly perspectival, fixing human creations in space relative to all other objects. In an attempt to locate the elusive soul and equally elusive consciousness, we have scanned, probed and analyzed the brain with increased zeal. Such endeavours are a type of 'shadow boxing before a mirror whose reflection occurs against the blind surface'. Almost unnoticed this ultra rationalization 'destroys the very thing achieved by authentic [understanding]: our abilities to gain insight into the psyche'.[47]

But, we cannot advance our understanding of consciousness by diminishing or denigrating the mental structure unduly. It is an integral part of the picture that we should try and 'see' in context — for what it is. As Edwin Abbot's book *Flatland* demonstrates, mental structure has in fact been very helpful in allowing us to conceptualize life in other dimensions. Abbot imagines four worlds: pointland, lineland, flatland and spaceland. It is interesting to notice the parallels with Gebser's zero, one, two and three-dimensional consciousness structures. In pointland, the sole inhabitant talks to himself, quite unaware that he is both speaker and listener in a closed world. In lineland, where inhabitants move like trains on a single rail, sound is all-important to warn one another of one's direction travel. In flatland, individuals can feel each other's shapes by nudging each other's edges in the 2D plane. Only in spaceland can individuals 'see' the shapes of their fellow inhabitants. So consciousness progresses from pointland, a world with no external sense, to worlds where hearing, feeling and sight increase the range of sensory perception.

Integral consciousness: a whole new language

Gebser's integral consciousness structure is perhaps difficult to understand for several reasons. Firstly, it is still incipient in the world: 'we can only hazard a cursory glance at ... initial indications which ... could lead to a new mutation'. Integral consciousness, according to Gebser, is arational (not rational or irrational). This observation parallels some tenets of transpersonal psychology which recognizes the shortcomings of mere rationalism, which it seeks to 'transcend'. This integration cannot be accomplished 'by the reactivationof those structures underlying it', or by regression to irrationality, but rather by 're-establishment of the inviolate and pristine state of the origin by incorporating the wealth of all subsequent achievement' in a 'fully completed and realized wholeness' in which 'the various structures that constitute [the individual] must become *transparent* and conscious to him'.[48] Gebser does not mean simply to shed the mental light of consciousness on the underlying structures — for that would be to destroy them. 'The true light illuminates nothing: it is itself everything.' Nor does he mean a mere expansion of consciousness (which is an obvious spatial concept). Rather he means an '*intensification* of consciousness' (his emphasis) and says that: 'pure transparency ... is understanding without an object, without a subject, a pure happening'.[49] We sense, feel, and know all at once. Perhaps this integral state could be equated with the notion of consciously reconnecting with the unconscious 'psychic web' of the biosphere, the collective unconscious, or what Fawcett calls the Universal Psychical Life. Gebser evidently had such experience which allowed him to refer to his early adulthood as 'the sleeping years' and contrast it with what Feuerstein described and called 'his *satori* experience'.[50]

Secondly, we have the problem of language. To represent the integral consciousness structure one cannot rely on the typical, object-specific perspective of the mental structure. As consciousness changes so does vocabulary: hence, Gebser's use of words like concretion, diaphaneity, sphericity and transparency. Some words like 'transparency' have already crept into our moral, ethical and political language, with meanings similar to Gebser (and 'diaphanous' more or less has a similar meaning). Others require more attention in order to understand Gebser's meaning. Here Georg Feuerstein's analysis of Gebser's work is helpful in including a succinct glossary. So for example, concretion, or what Gebser often refers to as time-concretion, is the direct concrete perception (as well as feeling, intuition and conception) of the 'spiritual presence', or the ever-present origin, characterized as the 'uncreated light' which 'is itself everything'. As Feuerstein points out, Gebser's 'reality appraisal' was 'unabashedly' spiritual. This simply means that he never shied away from the ultimate existential questions: 'Who am I? Whence do I come? Whither do I go?' These questions were always at the heart of his understanding of human culture and its many manifestations. Moreover, he clearly regarded scepticism and cynicism ('dis-ease') about such matters as a manifestation of the tendency towards spiritual discontent arising from the deficient mode of rational consciousness.[51]

One could, in theory, write or otherwise 'represent' each of Gebser's five consciousness structures in its own characteristic language (art form or musical genre), and indeed ancient and contemporary literature and art does just this for several structures. However, we would probably find some structures (for instance, the archaic) lacking anything we would recognize. As Owen Barfield often points out, 'meaning' changes fundamentally as consciousness evolves. Thus, our talk of insight into the 'big picture', or our growing perspective on the different 'dimensions' and structures of consciousness is riddled with the vocabulary of mental structure. Despite his originality, even Gebser was bound by the limitations of language.

Gebser was one of many twentieth century commentators to suggest that humanity was at a turning point. Our unease or 'anxiety' (see Chapter 2) was, in his opinion, a symptom of the breakdown of mental-rational order leading to apparent chaos. The phase shift from mental to integral consciousness, like any fundamental re-ordering of structure, creates a sense of upheaval or crisis. The Chinese characters for crisis translate as 'dangerous opportunity' — by definition a sense of unease, anxiety and chaos that cannot be rationalized by the discursive powers of reason. His suggestions now have the backing of the new science of chaos and complexity which admits ambiguity and uncertainty.

In contrast to some opinions and critiques discussed below, Gebser stressed that integral consciousness is not merely a synthesis of polar opposites A + B. This only leads us to run 'too and fro' at the mercy of dimensional constraints. The transformation is more like a leap, epiphany (see Chapter 3) or phase shift from a 3D world to one that is adimensional and aperspectival. Just as there is no well defined progressive direction to such transformation, so there is no timetable, 'the concretion of time is one of the preconditions for the integral structure'.[52] This is still hard to grasp for the mental-rational mind which experiences space and time as vaguely separate, abstract and measurable dimensions. But the challenges resemble those posed by modern physics which defines space and time as a continuum, that is still largely a conceptual abstraction and rarely a concrete experience (like precognition or remote viewing) that directly affects our actions in life.

Because our intellectual thinking is strongly conditioned by the mental structure, special effort, and some discomfort, is required to jolt or shift it. Gebser was sympathetic to all modes of consciousness, and it would be quite incorrect to suggest that he regarded an integral consciousness as an alternative to the mental structure. Nevertheless, integration is not achieved by thought alone. (One cannot observe one's own thinking except after the fact, and new thinking is required to analyze the previous thought just experienced. Old thoughts become the 'object' of attention by new thoughts and time remains linear and abstract.) In other words, time is not fully realized in us, for our thinking lacks intuition (literally 'inner presence'). Let us end with an easy-to-illustrate example of time integration. Gebser notes that the art of Pablo Picasso is non-dimensional in the sense that it allows us to see a subject, such as a human face or figure, from front, back and sides all at once. Such 'aperspectival' images jolt our normal sensibilities, in much the same way that the conundrums of relativity and quantum mechanics may, when they suggest that time and mass can dilate or shrink, that light may be both a wave and a particle or that the electron can be in two different locations simultaneously. Reality is far less measurable, and far more complex than we once thought. But, it is precisely the emergence of a complex, trans-rational, integral consciousness structure that is responsible for the new representations of the world that we create whether they be the paradoxes of the space-time continuum, the paintings of Picasso, or the notion that order arises from chaos.[53]

Repackaging Gebser: towards a grand integration

If you found any of the previous description of integral consciousness structure obscure or difficult to grasp, you are not alone. This is in part because it requires a different language, which is not always easy to express. However, failing to fully appreciate Gebser, may be partly due to shortcomings in *his* exposition, and its subsequent translation, or deficiencies in his interpretations. Both Georg Feuerstein and Allan Combs suggests that Gebser underplays the importance of the time line metaphor.[54] If Combs is right, we can fall back on the familiar, easy-to-understand linear model of history. This is essentially what Feuerstein and William Irwin Thompson did when they repackaged Gebser in a relatively simple historical framework as shown in Table 5.3.[55] In this scheme, approximate dates are put on the times in history and prehistory when various consciousness structures predominated. Feuerstein and Thompson both understood Gebser's reasons for avoiding such time scales, but nevertheless used a simplified scheme for ease of understanding. Thompson also integrated Marshall McLuhan's ideas on media systems to show how dominant means of communication have diversified through time, reaching ever-wider audiences with increased rapidity. Likewise the political structure or polity and means of cohesion has also changed through an expansion from bands and tribes, held together by rigid systems of dominance and authority, to larger city state, nation state and global entities administered by systems of justice, representation and participation.

Table 5.3. Gebser repackaged: modified from schemes proposed by Georg Feuerstein and William Irwin Thompson. Note that each 'successive' structure or phase allows for the integration of previous phases.

Consciousness structure	McLuhan's media systems	Social Structure	Polity	Cohesive association
Archaic	ORAL	Culture 200,000– 10,000 BC	Band	Dominance
Magic	SCRIPT	Society 10,000– 3,500 BC	Tribe	Authority
Mythical	ALPHABETIC	Civilization 3,500– 1,500	City State	Justice
Mental	PRINT	Industrialization 1500–1945	Nation State	Representation
Integral	ELECTRONIC	Planetization after 1945	Noetic Polity	Participation

Thompson built on Gebser's analyses of consciousness structures as they apply to individual psyches and worldviews, by showing how they manifest in the collective of human society. The historical process of individuation in the soul and Ego creates an interesting 'reverse' or reciprocal dynamic. Instead of individuals becoming ever more isolated 'islands unto themselves' they become ever more integrated into complex societies, where aspects of their individuality are at once valued for their diversity and constrained by and converted into the socio-cultural and political infrastructure. So, individuals, who were initially fully integrated within the unconscious, natural world, next become more psychologically (consciously) independent of one another, before finally re-integrated into new culturally created systems. Barfield calls this a cycle of original participation, separation and final participation.[56]

"	same as above
S-F	brain 'structure functions'
Conop	Piaget's Concrete Operation
Formop	Piaget's Formal Operation

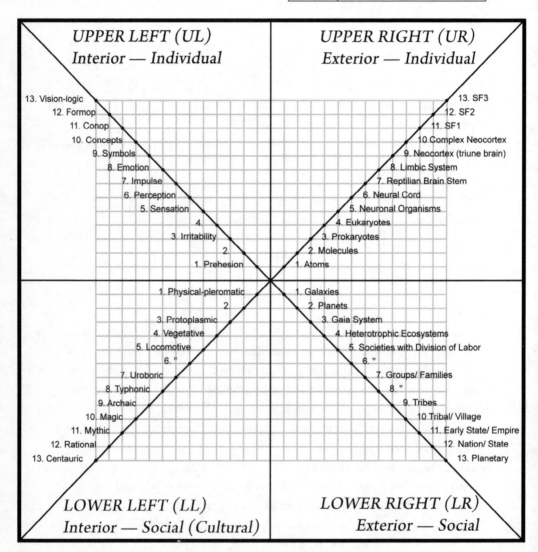

Figure 5.5. Wilber's four quadrant scheme. (Modified after Wilber 1998.)

Between Gebser, Feuerstein, Thompson and Combs, the evolution of consciousness is taken back to an early stage in human history. Their interpretations of when characteristic human faculties such as language and self-consciousness emerged are necessarily vague. Most authors agree that these are fundamental characteristics of *Homo sapiens*, whose origins go back at least one hundred thousand years. There is also agreement that representational art and refined 'late Paleolithic' craftsmanship both go back at least 30–35,000 years. The crude quality of tools and lack of significant art work, prior to this date are generally taken as evidence of more 'primitive' consciousness structures. For this reason, it is unlikely that anyone could make a case that pre-*sapiens* species manifest anything other than archaic consciousness, and we are sceptical of the idea that *Homo erectus* really 'understood' the golden ratio a million years before Pythagoras. However, there are many categories of sentience or sensitivity that can be attributed to pre-*sapiens*, hominids, primates, mammals and other animals: indeed to all life. But, one wonders if it is possible to define or describe such sentience in any meaningful way that gives us a sense of a continuous evolutionary complexification of the biosphere?

Ken Wilber has done precisely this. In an ambitious and now well known 'four quadrant' scheme.[57] (See Figure 5.5.) Wilber has shown how 'all' that we associate with our objective and subjective worlds, can be broken down into hierarchical arrangements in one of the four quadrants. As shown in Table 5.4, Wilber did not invent this scheme *de novo*. His framework reflects certain fundamental philosophical tenets discussed since the time of Plato.

So, the upper left quadrant represents our individual, interior (subjective), 'I' experience. The scale runs from the simplest level of sentience in organisms (prehension, irritability, sensation, and so on) up through the categories of emotional and cognitive development outlined by Piaget (see Chapter 3). The lower left quadrant represents the collective, interior (subjective) 'we' experience, and leads through collective categories of primitive life 'experience' (vegetative, locomotive, and so on) into Gebser's consciousness structures (archaic, magic, mythic, etc.). The two right hand quadrants both represent the exterior

Table 5.4. A simplified version of Ken Wilber's Four Quadrant scheme, with some of its ancient philosophical equivalents.

Wilbur's four quadrants		Philosophical-epistemological equivalents	
Upper left Interior— individual INTENTIONAL	Upper right Exterior — individual BEHAVIOURAL	I Beauty (Plato) *Art (Kant)*	It True (Plato) *Science (Kant)*
Lower left	Lower right	We	It
Interior— collective CULTURAL	Exterior — collective SOCIAL	Good (Plato) *Morals (Kant)*	True (Plato) *Science (Kant)*

world: the objective 'it' or 'other' categories. They are divided into a hierarchy of individual entities (atoms, molecules, primitive unicellular organs, and on 'up' to primitive and advanced brain systems) and collective entities (galaxies and planets from the inorganic world, and tribes, states and planetary organizations from the socio-organic world).

Wilber's model ambitiously attempts to integrate all objective, subjective, individual and collective components of the world/cosmos and, as noted below, many have argued that the integration is largely successful. A brief survey of the full quadrant model and its 52 categories (four hierarchical scales each with thirteen categories) shows that one can correlate from quadrant to quadrant, at least in principle. For example, the correlation of Gebser's consciousness structures with socio-political systems, such as tribes, nation states, and so on, closely mirrors Thompson's scheme (see Table 5.3). Likewise, we can correlate individual and collective entities in the external world (atoms and molecules make up galaxies and planets), or we can correlate individual organic complexity (cells, neurons and brains) with intentional behaviours (perceptions, impulses, and so on). Clearly the scheme is ambitious and open to scrutiny and potential criticism. Although we admire the scope of Wilber's integral scholarship, and avoid quibbling with the details, it is worth noting that the inherently grid-like Cartesian structure of the quadrant model is linear and hierarchical, and so symptomatic of the mental-rational paradigm. In fairness, the limitations of the quadrant's two-dimensional, map-like layout does not alter the seriousness of Wilber's claim that the model is the cornerstone of his 'Integral Theory of Consciousness'.[58]

The importance of Wilber's work is underscored by the attention it has received from a new generation of 'consciousness' scholars, including Allan Combs, who analyzed Wilber's work in great detail and Steve McIntosh, whose book is entitled *Integral Consciousness*.[59] Like Wilber, both these authors are aware that consciousness is a dynamic and highly complex phenomenon, and they have followed the evolution of Wilber's ideas through several iterations.[60]

McIntosh's Integral Consciousness model owes much to the groundwork of Gebser, Wilber and other consciousness gurus. Possibly the most striking and useful aspect of McIntosh's presentation is his use of a 'spiral dynamics' model (Figure 5.6) which is based on the work of the developmental psychologist Clare W. Graves.[61] Simply put, the ostensibly linear and progressive stages in development can be viewed as part of spiral dynamic which combines elements of both linear and cyclic time. McIntosh refers to this paradigm as exhibiting a 'conceptual and geometric elegance that clearly reflects evolution's dialectic method of transcendence and inclusion'.[62] This dialectical spiral of development resonates with previously articulated ideas about evolutionary progress which allow for the influence of new (transcendent or novel) and older (included, conservative) elements. Looking at the spiral model we see that McIntosh has identified eight consciousness structures: archaic, tribal, warrior, traditional, modernist, postmodern, integral and post-integral. These clearly reflect a modified Gebserian scheme, covering the archaic-integral spectrum. If we loosely substituted magic for

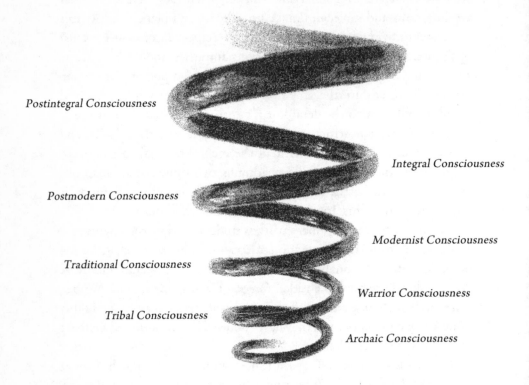

Postintegral Consciousness

Integral Consciousness

Postmodern Consciousness

Modernist Consciousness

Traditional Consciousness

Warrior Consciousness

Tribal Consciousness

Archaic Consciousness

Figure 5.6. The spiral model for the evolution of consciousness. (Modified after McIntosh 2007.)

tribal, mythic for warrior and traditional and mental for modern and postmodern we would more or less duplicate Gebser's scheme.

Here however, we are more interested in the dynamic relationships between structures than in defining precise, artificial labels or boundaries. The spiral's inherent dynamics are interesting. At any given place or time, society established its moods, norms and cultural conventions which we label as a 'developmental stage': its 'thesis'. But the dynamic flux of individual and collective consciousness soon leads to change, and reaction against the conventional norms or thesis creates the 'antithesis' as the next developmental stage. However, before long the antithesis becomes unsatisfactory and a new synthesis, combining elements of thesis and antithesis, emerges to 'complete' a cycle, but not in the sense that the cycles ever stop or end. The *thesis-antithesis-synthesis* dynamic — an idea derived from Georg Hegel's philosophy — is ongoing, and so structured that every developmental stage is at once a new thesis, antithesis of a previous stage and synthesis of two previous stages. This gives any developmental stage and its attendant consciousness structures a very rich and pluralistic or multicultural and synthetic flavour. However, it is worth reminding ourselves that Gebser specifically noted that integral consciousness was not a 'synthesis of any kind'.[63] The spiral as presented by McIntosh is reminiscent of the emotional-intellectual alternation of generations seen in the *faber-ludens* model and the alternation of individual and collective emphasis seen in the fourth turning model. It also reveals the dynamic organic tension that exists between novel and conservative forces.

When Gebser first wrote about 'integral consciousness', one wonders if he could have anticipated that, by the early twenty-first century, it would have matured into its own fledgling discipline, with departures which have modified his original meaning. According to McIntosh, the synthetic, integral thinking that points us in the direction of integral consciousness, or as Gebser might prefer, the sensibilities that arise from the eruption of integral consciousness, are glimpsed to various degrees in the work of a group of individuals (all men) who he identifies as integral philosophers: these include: Georg Hegel, Henri Bergson, Alfred North Whitehead, Teilhard de Chardin, Jean Gebser, James Mark Baldwin, Clare Graves, Jurgen Habermas and Ken Wilber.[64] Some may no doubt wish to add other pioneers to this list of integral philosophers, but as defined by McIntosh, the list suffices for now.

Nevertheless we should perhaps add the name of Giambattista Vico (1670–1744) to the list, if not as an integral philosopher, at least as an intellectual ground-breaker and founder of the science of sociology. Vico pioneered the writing of autobiography, publishing his own in 1725. Interestingly he wrote it in the third person, and on request, as an example to youth of an accomplished life. Of great interest here is his idea on the cyclic nature of history, proceeding from antiquity through the 'Age of Gods' to the 'Age of Heroes' and the 'Age of Men'. This cycle was subsequently repeated in more modern ages of gods, heroes and men represented by Christianity, the Middle ages and the Seventeenth century respectively.[65] Joseph McCabe, one of Vico's biographers, goes so far as to say it is obvious how readily Vico's 'doctrine could be adapted to the concept of progress as a spiral movement'.[66]

The best way to realize our interest in integral consciousness and the spiral of evolutionary development is to explore concrete rather than abstract examples that integrate all aspects of the experiential world. We attempt to do this in the next chapters by taking a deeper look at the body-mind as a dynamic and fully integrated manifestation of interwoven consciousness structures.

PART 3.

The Biological and Psychological Dynamics of Consciousness

6. A New Look at Who We Are and How We Got That Way

Integrating the cosmic individual

Psychology tells us that we are a complex mix of physical, emotional intellectual and spiritual desires. We have 'gut' feelings, 'heartfelt' emotions, 'brainstorms' and 'headstrong' tendencies, not to mention measurable brainwaves, intangible soul mates, spiritual yearnings, experiences and even 'guides'. Gebser went a step further in concretizing such notions when suggesting that the three M structures (magical, mythic and mental) correspond, respectively, to what we might call consciousness 'centres of gravity' in the gut, heart and brain (see Table 5.1 above). It certainly seems true that modern neuroscience believes that consciousness is centred in the brain, and we must take it that the Pueblo Indians genuinely believed this was all wrong when they told Carl Jung the centre was in the heart. In any event, Jung's holistic take on the body-mind accommodates a make up of willing, feeling and thinking.

If Gebser is correct, then centres of consciousness shift from gut to heart to brain, or at the very least, we humans perceive our consciousness to be centred in different parts of the body depending on our experience, age, culture and other factors. We may perceive that consciousness is everywhere — a type of universal primary datum that allows us to talk of a conscious universe.[1] Evidently, consciousness is a very variable phenomenon perceived to be anywhere, everywhere or nowhere in any given circumstance. Although talk of 'locating' consciousness 'anywhere' is problematic and very much a manifestation of mental thinking, we can only dismiss the idea that consciousness has a location, or many locations, if we are willing to tell neuroscientists, Pueblo Indians, Gebser and many others that they are deluded.[2] But this does not get rid of our tangible conviction that our consciousness is somehow a property of self, centred within us, and somehow also a property of sentient beings in the world we experience. Other alternatives are that it is somehow nonlocal and universal, like a force or subtle energy that binds the cosmos — the intelligence that so intrigued Einstein and stimulated his desire to know 'God's thoughts'.

In this chapter we shall continue to steer ourselves in the direction of an integrated, holistic understanding of consciousness by marshalling the scientific evidence for a deep integration between the body and mind. We will present biological evidence that there is 'a spectrum of consciousness'[3] deeply integrated or embedded in organic systems.[4] By spectrum we do not simply mean a linear scale of frequencies, as seen in a simple depiction of the electro-magnetic (EM) spectrum. Rather we prefer to suggest a plexus of simultaneous frequencies. For example, just as there are millions of coexisting individual humans, and an even greater multitude of species, so there are equally numerous manifestations of consciousness. At first such diversity might seem bewildering. However, just as it is possible to recognize patterns of organization in the biological world, with different species growing into different forms, and operating behaviourally at different frequencies, so we may recognize patterns of similarity and organization in consciousness structures. Here the fractal and recursion metaphors are useful: each individual is both generally similar, but different in specific characteristics of anatomy, physiology, behaviour and consciousness. Thus, any individual may have its characteristic spectrum of frequencies that correspond to age gender, species, health status and so on. We are complex nodes and fluxes deeply integrated in the psychic web of the biosphere, and part of the very organization of the cosmos. We are all invited to this 'happening' experience: there is no way out!

The EM spectrum flux of vibrations operates simultaneously at many different frequencies so that we can hear sound waves, feel infra red heat and see visible light all at once. Bees and dogs see and hear different frequency mixes from humans. Using a radio analogy, science already tells us that 'all' the frequencies or broadcasts are out there, whether they be neutrinos passing right through the earth, solar radiation giving us a tan, the gentle whisper of wind in the pines or the inaudible background whisper of the Big Bang, only picked up by cleverly-designed radio telescopes. We already know all these 'good vibrations' are out there and accept that, as individuals, we only consciously tune into a few, while 'below the radar' in the domain of biorhythms we tune in to many frequencies quite unconsciously. However, in understanding the natural, receptive ability of different individuals, species and instruments we happily acknowledge that we 'know' and 'verify' the existence of many signals not experienced directly or consciously. As a superorganism, the biosphere's collective receptivity is far greater than that of any individual. We are in this together, and rely on each other to understand those parts of the spectrum we cannot access directly. So, the ability of a dog to hear high frequencies is not doggy delusion, and we do not doubt the musician who has perfect pitch. For the same reason we should not dismiss the experience of individuals who report intensified or elevated states of consciousness, or the evidence that authentic mediums may be capable of 'tuning' into frequencies that are imperceptible to most of us. Thus, we take seriously researchers like Piaget and Gebser who have demonstrated changing consciousness structures in child development and throughout human cultural history. Their ability to tune in to the broader collective consciousness message is all part of the dynamic process which allows us to know ourselves individually and collectively and find our centres within the larger 'integral' sphere.

In this section we shall tease apart the patterns of organization that make up our cells, bodies and our biosphere, and show how they affect the birds, bees and mammals equally, and most profoundly. This in itself is a shift in consciousness from anthropomorphism to a type of mammalo-morphism or biosphere-morphism. For good measure we shall also try and jettison the idea that linear time is marching evolution down the line from past to present, from primitive to advanced. As modern physics now knows, time is not a backdrop, a scale against which the evolutionary process runs like an Olympic race. The competitive evolutionary 'arms race' metaphor has some validity, but it is only part of the story. As noted below, competition fails to take account of symbiosis, ecological interdependency and other co-operative dynamics essential to organic systems.

We are at a stage or 'turning point' in the evolution of consciousness when, as Gebser predicted, integration is inevitable. It is manifest all around us in media, science, medicine and culture. So, our new task is to move beyond theoretical expositions to show how a new integral consciousness structure can enliven the way we view our very own biological organization, and give insight into how we view our bodies, and those of our fellow species, as vessels of evolving and 'understandable' consciousness. In short, conscious evolution is penetrating and permeating physics, biology and psychology, helping us better know ourselves and ultimately our place in the cosmos.

Reading the dynamic organization of the biosphere

We may see the biosphere as a tangled chaos of organisms, a buzzing confusion of species struggling to survive. But even though plants don't grow in linear rows without the aid of farmers, life in the biosphere is far more ordered than we might think at first glance. Animals and birds may appear to roam at random, but in fact they have well defined niches and roles as part of an integrated dynamic process of organization. Indeed the very definition of organization is a patterned relationship between organs or parts. Clearly, the interdependence of processes such as atmospheric circulation, carbon and nitrogen cycles and plant and animal respiration indicate large scale patterns of biological organization. Indeed this is the main message of Lovelock's Gaia Hypothesis: Earth is a superorganism.[5]

Biological organization is not just about food chains and species interactions described in basic ecology texts. Popular talk of 'deep ecology' indicates profoundly mysterious, almost spiritual connections that we intuit subconsciously but do not fully understand.[6] One such deep and well studied connection is manifest in our genetic makeup which clearly shows how all Earth's species share a common biochemistry and a common origin. Other recent studies show that organisms are organized 'coherently' as so-called molecular democracies — energetically integrated like magnetic fields.[7] Each organism is, among other things, its own electromagnetic field, like a microcosm of earth itself. So we are fields within fields. Just as consciousness manifests coherence between gut, heart and brain organization, so there is biosphere-wide, electromagnetic, biochemical, and genetic coherence.

How do we understand this organization? How do humans, cats, dogs or plants each evolve distinct form (anatomy), behaviour and consciousness? Why are humans rather like apes, and cats so much like tigers? Why do we simultaneously see shades of unity and diversity? Can we address Darwin's unanswered question on the origin of species? What 'process' gives rise to these behaviours and appearances that we call characters or characteristics?

We believe that the answer is simpler than one might anticipate, even if only clearly stated in some circles. Organisms seem to obey some very definite laws of growth and behaviour, which biologists often discuss as form and function. Patterns repeat endlessly, but fractally with a little novelty each time. These creative processes are fascinating and understandable even to the non-biologist. What is more they help us see behaviour and consciousness in a new light. Science likes this. The simpler elegant answers are preferred and help us unify otherwise disparate threads.[8]

If we begin with the biosphere, we find it is organized as a superorganism comprised of many smaller units or organs. The biosphere is huge, covering 510 million square km of the Earth's surface. These organizational or structural dimensions extend for many orders of magnitude above and below the scale of any individual organism such as a human being (see Table 6.1, and Figure 1.1).

The biosphere also operates on evolutionary time scales that are both longer and shorter than our experience, either as a one hundred thousand year old species or as individuals who might just live one hundred years. Intriguingly, as suggested in Chapter 1, we seem to be right in the middle of the spatial and temporal spectrum — quite probably because this is as far 'up' and 'down' as our consciousness allows us to perceive.

Table 6.1. The organization of the biosphere.

Biological order of magnitude	Biological entity	Lifespan
10,000–100,000 km	Biosphere	3.5 billion years
1,000–10,000 km	Biosphere region	Millions of years
100–1,000 km	Large ecosystem (reef)	Millions of years
10–100 km	Intermediate ecosystem	
1–10 km	Community — ecosystem	
100 m–1 km	Population — species community	
10 m–100 m	Large vertebrate	80 years
1 m–10 m	Human —large vertebrate	
10 cm–1 m	Organs, small vertebrates — large invertebrate	
1 cm–10 cm	Tissue systems, chicken egg — invertebrate	
1 mm–1 cm	Large cell — frog egg	
100 μm–1 mm	Large eukaryotic cell	
10 μm–100 μm	Eukaryotic cell	20 minutes
1 μm–10 μm	Bacterium — mitochondria	
100 nm–1 μm	Virus — bacterium	
10 nm–100 nm	Ribosome — virus	
1 nm–10 nm	Protein molecule	
0.1 nm–1 nm	Atoms — small molecules	

To prove the biosphere is coherently organized throughout, we shall begin with a simple exercise in logic. If we ask whether the biosphere is coherently organized at levels where it has been studied intensively: that is, at the level of the cell, the organism, the species and the whole Gaia system. The answer is yes, in all cases. The cell is in many cases

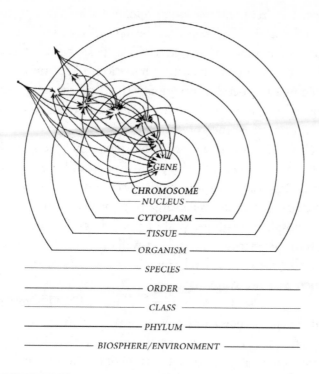

GENE
CHROMOSOME
NUCLEUS
CYTOPLASM
TISSUE
ORGANISM
SPECIES
ORDER
CLASS
PHYLUM
BIOSPHERE/ENVIRONMENT

Organic Unit	Size (spatial dimension)	Life span (temporal units)
Individual animal	1–10 tons biomass	10–100 years
Individual species/genus	Single population	7 million years
Brachiosaurs	20 species low latitude	100 million years
Dinosaurs (sub class)	800 species low-mid lat.	150 million years
Reptiles (class)	6,000 species low-mid lat.	350 million years
Vertebrates	ca. 50,000 global range	550 million years
Biosphere	Global surface (all species)	3.5 billion years

Figure 6.1. Complex relationships in hierarchical biological systems (after Weiss 1973), with analogous temporal hierarchy. (After Lockley 2008.)

a unique organism, whether it is a pre-nuclear (prokaryote) organism like a bacterium, or a nucleus bearing (eukaryote) like an amoeba, or even a giant strand of seaweed (kelp). Cell functions maintain homeostasis: that is, a balance of regulatory process, such as respiration, ingestion and excretion. Likewise the individual organism operates independently from other organisms on many levels, retaining integrity through homeostasis. The same can be said of the whole biosphere. This is the main tenet of the Gaia hypothesis which regards the biosphere as a superorganism, with circulation, respiration and other dynamic biological processes. Few doubt this, and the only significant debate centres on whether the whole system is conscious, and how such consciousness can be defined or understood.[9]

Given this cell-individual-species-biosphere hierarchy it is only logical to ask whether all these entities could be functioning homeostatic organisms, without similar homeostatic organization at the intermediate levels of tissues, organs, populations and ecosystems. Could the biosphere really be well organized at some levels and not others, or have alternating well and less-well organized levels? A related question is: how do these levels interact? Students of such complex systems such as Paul Weiss have argued that all levels are linked in complex and manifold ways.[10] (See Figure 6.1.) The relationship between different domains or levels of organization can be explained using the concept of the holon introduced by Arthur Koestler.[11] The holon is a coherent organizational entity composed of 'smaller' often less complex holons, and nested within 'larger', often more complex holons. Thus, the cell is a holon within an organ, which in turn forms part of an organism. Each is its own entity, but each is nested like a Russian doll in a larger entity. (See Figure 6.2.)

Holons teach us something important about the organization of systems. Every system is both open and closed at the same time. We humans are perfect examples. We have to be open to the environment in order to eat, breathe and communicate with others. But at the same time we do not dissolve in the rain or evaporate into thin air. Our skin serves as a physical but permeable boundary between us and the environment. It is much like the membranes that separate cells without isolating or closing them off. In psychological terms, we can be thin-skinned and too susceptible to environmental stimuli, or thick-skinned and more autonomous. If we were completely open or completely closed systems we would perish from overstimulation, starvation or asphyxiation. Koestler also recognized that this balance between being open and closed was a result of the interplay of forces of integration and

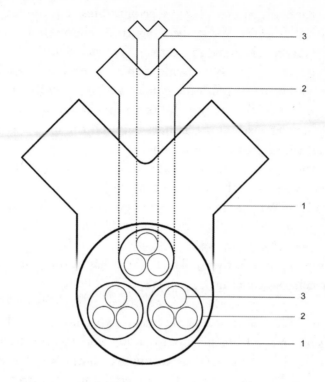

Figure 6.2. Diagrammatic depiction of holons. (Modified after Koestler 1969.)
The numbers show diminishing levels of magnitude.

disintegration.[12] This same polarity principle was expressed by Douglas Fawcett's ideas on the balance between novel and conservative forces that open or close us to various degrees. (See Chapter 2 above.)

There are many classic examples of biological organization. The hive of bees or colony of termites is a collection of individuals acting as a superorganism. Likewise the flock of 10,000 birds or the shoal of a million fish, that all turn in the same millisecond without a signal from any leader, seem to have some uncanny psychic or telepathic communication organizing their movements. (Incidentally, in movie scenarios, most mind control victims are unconscious.) Most studies of social insects characterize individuals both as organisms and cells or units that make up much larger superorganisms. The bee is not really independent of the hive. Worker bees and termites belong to a caste system dependent on the colony and they cannot reproduce on their own.

Likewise, almost every individual creature on Earth has its own distinct masculine or feminine gender and so is really incomplete and dependent on the other sex in order for the species to be whole. In many species distinct differences between the male and female, known as dimorphism (two morphologies) are the rule rather than the exception. It is in our language: 'as a rule', we say nature gives larger horns to the male ungulate than to the female, and no horns to the rodent. Why is there a two-fold gender rule? Why not three, four, six or ten sexes? No expert on morphology, biochemistry or genetics has a simple answer.

In insect colonies or the coral reefs, there may be more than just two types of 'cell' or gender making up a species. The termite colony caste system has five types: workers, soldier, winged reproductives, kings and queens. Is this an exception to the rule of two genders? The honeybee hive has a threefold organization: queens, workers and drones. (See Figure 6.3.) Each has its own distinctive morphology, which can be understood as a variation of a universal pattern. Insects such as stag beetles show striking morphological convergence with quite different species like deer (stags and does). But do they converge by

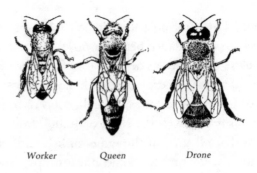

Worker   Queen   Drone

Figure 6.3. Threefold morphology in bees.

chance or are they evidence of an underlying pattern of organization? Leading biologists have pondered this question by asking whether things would necessarily come out the same again if the evolutionary tape of life were rewound and replayed. One camp, led by Stephen J. Gould (the late Harvard biologist from Cambridge, Massachusetts), said no, it would all be different. The other camp, led by Simon Conway-Morris (Cambridge University, England) says yes, the same basic patterning is deeply engrained in the fabric of organic process.[13] As we shall show, there is supporting evidence for the latter view.

But first we must take a journey in time (and space). All things grow and develop. We assign timetables to such processes whether they are the 3.5-billion-year-long evolution of the biosphere, or the short hour of flight experienced by the ephemeral mayfly. But as we know these are more than simple timetables like train schedules or the daily nine-to-five routine. Time as an 'agent' is far more complex and infinitely more dynamic.

Evolutionary timescapes: biological clocks and heterochrony

How long do we live? For the average individual human the answer is about three score years and ten. Seventy is a nice round number, although insurance companies are more statistically precise depending on your sex, nationality, health habits and other factors. They will also tell you that this average is changing, and your chances of long life are increasing.

However, when asked in a broader context, the question becomes far more complex. If we are talking about the longevity of our families, we may be able to trace direct relatives back for hundreds of years, and our species has been around for about 100,000 years. Paleontologist trace our primate order back some 65 million years, to the last days of the dinosaurs, and they trace our mammal line back 300 million years to the age of Carboniferous coal swamps. We can trace fish and backboned animals back to about 540 million years, or go the whole hog and track life back some 3.5 million years.

Sticklers may consider this cheating, but, must we stick to the definition of the life of individuals? How do we define the lifespan of a bee? By the worker, the queen or the hive? How old is a coral reef?[14] Or better still, how old is the asexual bacterium that clones itself so precisely that it is effectively immortal from generation to generation.[15] Among living species, the bristlecone pine lives for thousands of years, the microbe for mere minutes. Why? Why does the mouse live a few years at most and the tortoise, elephant and human for a century or more? An even more intriguing question is: why do both mouse and elephant have the same average number of breaths (about 190 million) and heartbeats (4 per breath) in a lifetime but just experience them at different rates?[16] The great god of cosmic organization somehow gave almost all mammals the same allotment of breaths and heartbeats; it is just that the mouse literally vibrates at a much higher frequency than the elephant. The human is the main exception to the rule with 660 million breaths, although other primates have also been given above average allotments.[17]

So time apparently shows favouritism to some species in the longevity stakes, but only if you use our current human measures of time. Ironically, humans have most-favoured status, but this good news is obviated by being aware of our mortality. Thus, the bad news is that we are haunted by time. Measured in mammalian heartbeat units the time gods are more even-handed. But differences are not just seen in comparing large and small species. If we look at our own cells we find that some (our brain cells) last a lifetime, while others (our stomach cells) are being replaced by the minute. Our heart beats about 70 times a minute, but we breathe only once for every four heartbeats. And, of course the rate changes, depending on how fast we move, both physically and emotionally. These different rates of physiological activity have been referred to using the metaphor of biological clocks. There is now a whole science of chronobiology (literally time biology), that measures, not just physiological activity in the body (Figure 6.4), but the complex relationship between these 'internal' processes, and the external process such as lunar cycles, which affect the reproduction of thousands of species including our own. In the short term we known that our internal rhythms

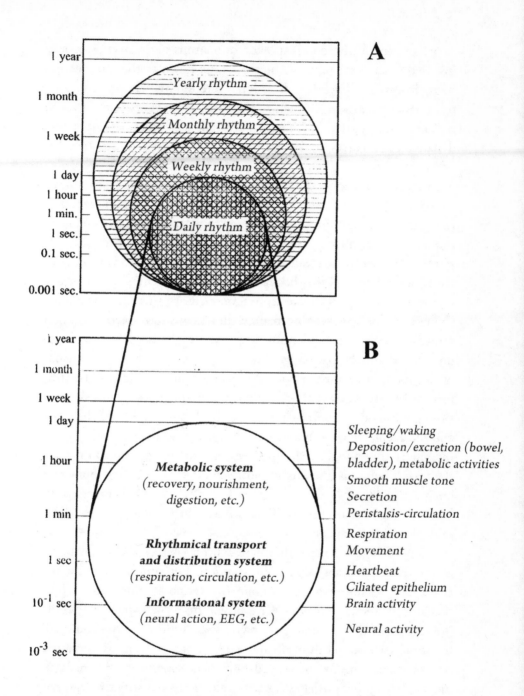

Figure 6.4. Chronobiological rhythms affecting humans. A: daily through annual cycles. B: daily through millisecond cycles. (Modified after Rohen 2007.)

vary from the milliseconds it takes for neurons to fire, to the seconds involved in the heartbeat and breathing, on down to the minutes and hours involved in cycles of digestion, excretion and sleep.[18] Our long term development cycles, such as the three seven-year cycles of tooth replacement discussed in Chapter 4 are tied to complex developmental dynamics, that correspond to well recognized changes in the structure of our consciousness.

In order to understand better these changes in consciousness from a biological perspective, we should appreciate another dynamic aspect of time that is already quite familiar to us. This is the unexpected quantum leap or wrinkle in time rather than the steady march. Chronobiology mostly deals with the steady beat of the drum, the steady ticking of the biological clock. Like the seasonal or lunar rhythms, that regulate reproductive cycles, or the steady beat of the heart, there are few sudden changes in the timekeeping regime. But as we have seen, evolution sometimes does make quantum leaps, and biologists have sought to understand this by invoking different timing processes. The science of heterochrony, which means 'different timing', addresses just such processes.[19] In the traditional language of biology, we can ask what timing mechanism controls the eruption of baby teeth and their loss and replacement by permanent teeth. Of extraordinary evolutionary interest is the fact that chimpanzees and other apes are born with their teeth, whereas humans are not. Something in the developmental 'program' evidently delays the growth of teeth, and other organs, in humans but not in chimps. This differential timing is an excellent example of heterochrony.

Many biologists believe that heterochrony is responsible for many of the quantum leaps we see in evolution. Put simply, some believe that fish, early representatives of our own vertebrate line, evolved by a process of heterochrony which freed them from the sedentary lifestyle of a group of their ancestors known as sea squirts.[20] One could fill libraries with the technical, and still under-appreciated literature on how heterochrony influenced human development and made us distinct from other primates. Virtually all our most distinctive features, including nakedness, flat faces without protruding jaws, and even upright posture can be more or less explained by this process.

Despite the importance of heterochrony, until recently the subject has been largely ignored by most biologists. The aforementioned Stephen J. Gould was an exception. His classic book *Ontogeny and Phylogeny*, outlined the connection between the processes manifest in early or late growth of organs, such as teeth, during individual development (ontogeny), and the analogous large scale processes that probably caused sudden quantum leaps during long term evolution (phylogeny).[21] Thus, Gould took a modern (1970s) look at a subject that has intrigued and perplexed biologists since before the 1860s, when the term heterochrony was first coined by the German biologist Ernst Haeckel.[22] Haeckel, who did as much to promote evolutionary theory as Darwin, Wallace and Huxley, proposed the famous 'biogenetic law' which suggests that ontogeny recapitulates phylogeny. This law appears to explain how development (ontogeny) from a single, fertilized cell to fully formed human, proceeds through stages that resemble our entire evolutionary history (phylogeny) through fish- and amphibian-like stages. Ostensibly our development is like a miniature, microcosmic replay of all of evolution.

Although at first rejected by many biologists on the grounds that exact replays would only create identical clones, in recent years recapitulation has come back into vogue, and Haeckel certainly deserves much credit for popularizing the concept of evolution. We now recognize that development is a type of recursion, but not an exact replay. Each time around, with each new life, and with each new species, there is some novel variation: something new. *'Vive la différence.'* This is the fractal principle of recursion in action. Heterochrony is the differential timing agent that intervenes to reset the developmental timetable. So in humans, for example, childhood is prolonged, this makes us helpless at birth and dependent on parents for many years, but this shift allows for prolonged periods of play and learning which clearly has a huge impact on individual and social consciousness.

Heterochronic shifts probably account for differences in brain size and heartbeat allotments between humans, higher primates and other mammals. For example, as all higher mammals have large brains relative to their body size in early fetal development, the simple process of 'retarding' development in a more juvenile stage will result in a relatively large brain compared to the size of the rest of the body. Other fundamental shifts have taken place in the timing of puberty. As noted in Chapter 3, even in historical times humans have reached sexual maturity (biological adulthood) earlier and earlier. At the same time, humans are the only mammals to have a long, post-reproductive phase of life. These heterochronic shifts result in pronounced changes in anatomy and physiology while also creating novel possibilities for the development of behaviour and consciousness. Only a few researchers, such as Rudolf Steiner, Wolfgang Schad, Johannes Rohen and Jos Verhulst have investigated these changes comprehensively and understood their far-reaching implications.

As noted in Chapter 3, Steiner understood the educational implications of heterochrony and founded the Waldorf Educational system to facilitate harmonious resonance between the physical, emotional, mental and spiritual development of children in their formative years. While justly recognized for his pedagogy, Steiner's other contributions are foundational to the biological paradigm presented here. Steiner comprehensively interpreted the scientific work of Johann Wolfgang von Goethe (1749–1832) who, despite being better known for his literary work, was a pioneer biologist who founded the science of botany, made landmark discoveries in human anatomy when he found the human intermaxillary bone, and was the first to coin the word 'morphology'.[23] Goethe used the term morphology in a highly dynamic sense, to mean change in form during growth or development. Thus, it meant something like 'metamorphosis' or 'transmutation', both of which had the connotation of what today we might call evolution. Indeed the use of the word 'mutation' in modern genetics is something of a holdover from pre-Darwinian, proto-evolutionary vocabulary. At this time, and even to this day, the strange and unpredictable growth spurts of certain plants were called 'sports' (not spurts), or 'mutants'.

Goethe, also made another very important observation, that in our opinion is fundamental to understanding the dynamics of development in organisms. He recognized that it was impossible to change the morphology of any organ without an effect on adjacent organs, a concept he called the 'compensation principle'.[24] It was not until comparatively recently that modern students of heterochrony independently 'rediscovered' this compensation principle and labelled it a 'trade off'.[25]

Wolfgang Schad, a contemporary German biologist has developed the Goethean tradition into the comprehensive synthesis outlined below, revealing, as we hope to show, how compatible it is with the superorganism-complex system paradigm now emerging into scientific consciousness. In our opinion, Schad's insights amount to a new and revolutionary, integral theory of evolution.[26] Likewise, in a landmark book on *The Dynamics of Human Development,* the Belgian theoretical chemist Jos Verhulst promulgates what is essentially a new theory of evolution referred to as 'synergistic composition' which explains the importance of changes in the duration of infant, adult and post-reproductive phases of development in augmenting human potential.[27] In the sections that follow, we demonstrate that since the days of Goethe the shadow of a new evolutionary theory has been lurking in biological consciousness, waiting for the time to emerge into full awareness. That time, we believe, is now.

The organic Goethean-Schadian paradigm presented in the following sections does much to integrate all essential biological knowledge. It is a highly holistic tradition whose origins predate Darwinism. While too much ahead of its time in pre-Darwinian generations, and only partially realized by Germans like Haeckel, the Goethean paradigm is now enjoying renewed attention in the current biological revival known as the 'evo-devo' (evolution of development) movement, that for all its new jargon is basically a re-packaging of many of the tenets of heterochrony. The new paradigm does not throw out Darwinism. Rather it integrates much of what both modern and ancient science and philosophy have brought to light over many centuries. In the spirit of Gebserian integration, it is not a chronologically chauvinistic celebration of the latest discoveries, rather it is a demonstration that true perennial wisdom is timeless, and always richly fruitful.

How organisms take shape: the dynamics of head-to-toe growth

From head to tail, from head to toe, from top to bottom or from back to front: no matter how we express it we seem to have an innate sense of our orientation in space, and this sense of direction is deeply embedded in language. Head to tail orientation is common to all vertebrates and many invertebrates, and is thus a near universal characteristic of animal anatomy. Most molecules, cells, whole organisms and superorganisms grow by adding tissue and body mass in a systematic way from head to toe and from toe to head. Others add tissue in the 'middle' but then still shunt it in the direction of the head or tail. Anterior to posterior or cephalo-caudal growth, which in both cases quite literally means from head to tail is well known in biology.[28] We humans provide excellent examples. We are born with very large heads, relative to our feeble limbs, but as we grow, through childhood and adolescence, we grow into our bodies, and especially into our long limbs. (See Figure 6.5.)

For every action there is an equal and opposite reaction. So, in addition to head to tail trends we find the opposite trend of tail to head or posterior-anterior (caudo-cephalic), growth. During vertebrate evolution there is clear evidence that animals developed progressively larger brains over hundreds of millions of years. This evolutionary shift, therefore, proceeded from tail to head. As is well known the tail was quite literally lost in many lineages including frogs, birds and great apes. Interestingly, our friend Gebser reported shifts in consciousness from posterior to anterior organs. Such shifts are a type of dynamic developmental wave that that proceeds from one end of the body to the other and back again. Armed with such information, we are now prepared to understand that these developmental waves, work at different levels and on different timescales, starting with microscopic, short term molecular dynamics, and moving up the biosphere scale of space and time to the broad sweep of vertebrate evolution.

As the twenty-first century dawned, biology contrived to place ever-increasing emphasis on the molecular and genetic basis of life. This has led to the need to reinvestigate the complex dynamics of the

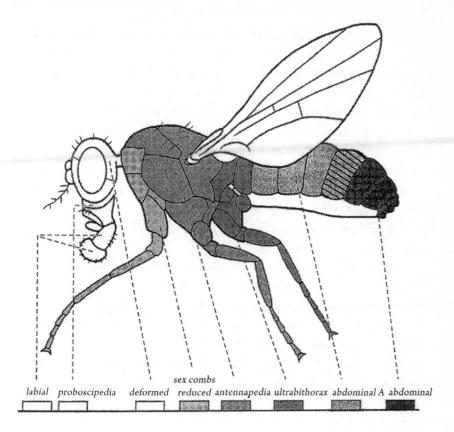

Figure 6.5. The anterior-posterior (A-P) organization of Hox genes (in the homeobox)
demonstrates strong parallelism with the A-P organization of actual body organs.
(Modified after Carroll 2005 and Lockley et al. 2008.)

recently dubbed 'evo-devo'. In the early day of genetics, the relationship
between genes and characters such as eye and hair colour were thought
to be quite simple: that is, a given gene determines a given character.
It has since become clear that the situation is far more complex:
single genes influence many characters and many genes affect single
characters. However, out of this seeming complexity some remarkable
correspondences or convergences between the microscopic molecular
world of genes and the macroscopic world of visible morphology, have
been recognized. The key to this understanding has been the recognition

(or more precisely the biological definition) of the Homeobox. This concept recognizes that the genes associated with the anterior and posterior organs of such diverse organisms as mice and fruit flies have the same anterior-posterior positions as their corresponding organs.[29] In other words, the head to tail organization is similar, parallel or convergent in the microcosm and macrocosm. (See Figure 6.6.) Indeed the term Homeobox could be translated as the 'same block' or the 'same' organizational pattern. Given this evidence of 'sameness', it is easy to see how the development, activation or 'expression' of Homeobox genes, known as Hox genes, follows an organized anterior to posterior sequence corresponding to the anterior-posterior development of tissues and organs in the body as a whole.

The same organizational principles apply to back-front (or dorsal ventral) organization. In 1822, in an extraordinary case of prescience, the holistically-minded French biologist Geoffroy Saint-Hilaire, claimed that invertebrates (arthropods) were really just vertebrates turned upside down.[30] Saint-Hilaire was a fan of Goethe's holistic thinking and its inherent compensation principles. However, his compensatory dorsal-ventral orientation ideas were shot down as nonsense by his erudite rival Baron George Cuvier, a famous pre-Darwinian who believed in the 'fixity of species' rather than in evolutionary change. It was not until the 1980s and 1990s that geneticists recognized the fact that the same genes that code for dorsal organs in invertebrates code for ventral organs in vertebrates and vice versa.[31] Suddenly, authors of specialized genetics papers, otherwise devoid of pre-1980s citations were quoting a forgotten 1822 article by Saint-Hilaire. In a tribute to the Frenchman's insight, Gould stated that 'the art of finding timeless essences in apparent trifles is the kind of perception that we call genius'.[32] This is a classic example of how nature's timeless wisdom is deeply embedded in subtle patterns of organization. Perhaps the new paradigm we advocate here pays greater credence to such perceptive subtlety.

It may seem like an obvious or intelligent economy of design, to organize the micro- and macrocosm along the same lines, but the mystery is how such organization comes about. Evo-devo studies tell

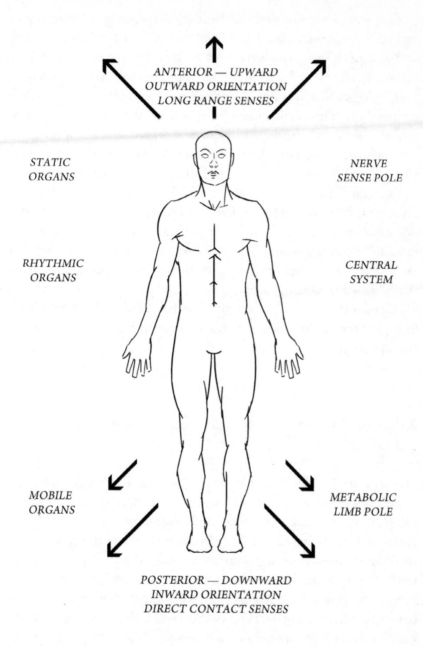

Figure 6.6. Anterior-posterior polarities in humans. (Based on the work of Wolfgang Schad 1977, and modified after illustrations by Lockley 1999, 2008.)

us Hox genes follow timetables, effectively 'expressing' or switching on and off from head to toe, sometimes starting a little earlier or latter, and at other times expressing more to the anterior or the posterior. Such differences probably play vital roles in creating the variety of morphologies that distinguish individuals and species, and probably control how individuals of the same species, with otherwise similar growth programs, develop different sex organs at critical stages of development.[33] So, Hox gene studies have recently claimed to hold the key to understanding some fundamental aspects of morphological organization and its expression in biological evolution. However, as the Saint-Hilaire example shows, important dynamics of bodily organization were already recognized long before the corresponding head to tail organization of Homeobox genes was mapped out. We must give credit where it is due. In doing this we cannot overlook the work of Wolfgang Schad, whose innovative, holistic work on man and mammals delves deeply into the fundamentals of bodily, biological integration and its expression in morphology, behaviour and consciousness.

Knowing ourselves: new perspectives on humans and mammals

Humans are mammals and so related to familiar species like cats, dogs, horses and cows, not to mention our close primate relatives, the monkeys and apes. But how has evolution produced such diverse forms as mice and men? We have already seen that both the genes and their anterior-posterior organization are quite similar even in animals that are very different in other respects (mice and fruit flies). The innovative approach of Wolfgang Schad was not to look at the genes but to examine the organization of the whole animal and especially its major organs. Is the mouse nervous because of an overactive nervous system? Does the cow digest tough cellulose because of an over-developed powerful digestive system common to all ruminants? Does the difference between nervousness (neurosis) and placid rumination tell us something about mammalian consciousness or indeed about our own human consciousness? As we shall see, the answer in all cases is probably yes.

In his book *Man and Mammals: Towards a Biology of Form,* Schad offers us an entirely new perspective on mammals.[34] This perspective allows us to see ourselves as we have never done before, and we venture to suggest that it is a deeply satisfying and intuitively and holistically correct view for anyone with a modest understanding of animal biology and behaviour. Schad reveals human organization as a polarity between the open anterior pole with its long-range senses (especially sight and hearing) situated in close association with the nervous system (brain), and the closed posterior pole of the digestive system and limbs. In between the two, the central system of heart and lungs (circulation and respiration) opens and closes rhythmically. (See Figure 6.7.)

This organization requires further explanation in order to be fully appreciated. A little reflection reveals what is meant by the open and closed ends of the system. When the human, like any other mammal, wakes up, its senses, particularly sight and hearing, are acutely sensitive or open to the external environment, including things that can be seen and heard at great distances. The anteriorly-situated brain, the centre of the nervous system, is in a state of waking consciousness. By contrast the posterior pole of limbs and digestive system are comparatively 'closed' to the external environment. The stomach deals only with what is inside it, and the limbs, which are blind and deaf, feel only the things with which they are in direct contact. These posterior systems are in a different state of consciousness more akin to sleep or the unconscious. For example, the stomach digests food in much the same way whether the organism is asleep or awake, and the limbs may function in the same way, as proved by sleepwalking, The middle system of heart and lungs mediates between the two poles of alert waking consciousness and unconscious sleep. It is the realm of subconscious dreams and emotions.

These simple examples establish some important connections between major organs and consciousness, which help us more fully understand the biological connections between consciousness and anatomy in man, mammals and other organisms. Let us begin by looking at the middle system and its mediating role between the

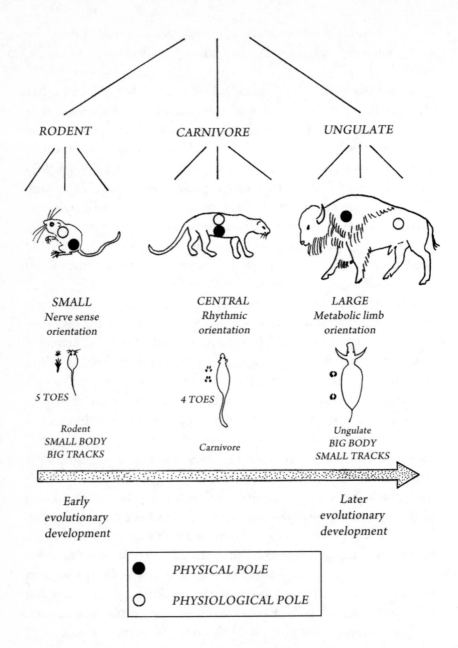

Figure 6.7. Anterior-posterior polarities in three main groups of placental mammals. (Based on the work of Wolfgang Schad 1977, and modified after illustrations by Lockley 1999.)

anterior (open) and posterior (closed) poles. In human medical and cultural history, the middle system plays a special role in defining life. An individual is considered alive as long as they are breathing and the heart is still beating. Thus, the middle system functions by constantly switching on and off, opening and closing, as we breathe in and out and open and close our heart valves. This rhythmic dynamic recurs throughout the whole body as it cycles between open wakefulness and the closed state of sleep. Even the *faber-ludens* cycle is something of a rhythmic alternation between being open or closed to intellectual or emotional activity.

These dynamic organic polarities repeat over and over again in a recursive pattern of organization that is easy to 'see' directly in morphology. For example, the anterior head region is 'outwardly' oriented or 'open' with external sense organs, whereas the posterior gut is 'inwardly' oriented or 'closed'. Thus, it is no coincidence that the skeleton is on the outside in the head enclosing the 'flesh' of the brain whereas in the limbs and posterior regions of the body, the flesh is found on the outside enclosing the bony skeleton. The middle region of trunk or thorax is more enclosed anteriorly than posteriorly and the ribs and abdominal muscles form a very obvious rhythmic pattern of segmentation.[35]

These striking examples of inward and outward organization along the body's anterior-posterior axis are only one expression of dynamic patterning. If we look at the anterior organization of the sensory system and the nervous system, we see that the senses are the outward component and the nervous system the inward, physical organ which 'digests' the external sensory information (which in the form of sight and sound waves is essentially intangible). Likewise the lungs are the outwardly connected part of the middle system which brings in invisible air to be digested in the visible blood, and the limbs are the outward part of the limb-digestive system which brings food, via the mouth, into the stomach. In hunting and trips to the restaurant, the limbs also bring the stomach to the food. As shown in Table 6.2, the whole body outward-inward polarity, which corresponds to anterior-posterior orientation, is repeated within each of the three systems: anterior, middle and posterior.

Table 6.2. *The anterior-posterior, outward (open)- inward (closed) organization of the major organ systems in humans (modified after the work Wolfgang Schad). Note the whole body outward-inward polarity (top to bottom) is repeated (and depicted from left to right) at the threefold level of the anterior, middle and posterior organ systems.*

Outward		
sense	speech	nerve
OUTWARD	respiration · circulation	INWARD
limbs	reproduction	metabolic (digestion)
Inward		

The human or mammal head is a microcosm of the body with the outwardly oriented eyes (and ears) more anteriorly located than the nose which is anatomically central and connected to the middle system. The mouth and jaws are posteriorly located, connected to the stomach, and like the limbs, they are mobile. They are the limbs of the head and serve as a means of picking up food in animals that cannot do so with their limbs. The recursive or fractal patterns of organization run much deeper at the levels of tissue. For example, ecto-, meso-, and endo-derm (that is, outer, middle, and inner layers of skin) are precisely the respective layers that give rise to the anterior, middle and posterior (or outer, middle, and inner oriented) organs.

Schad noted that if we look closely at the anterior system, we will see that it also contains the speech system as its own middle or mediating system, which also has a central location between the anterior and middle system. Speech is highly rhythmic: vowels are open and consonants tend to be closed, and syllables, words and phrases all represent multi-layered patterns of alternation between the sensory system and the nervous system in a constant process of feedback. For example, in a simple statement like, 'Look out! — No, it's ok,' speech has conveyed the idea of possible danger to the brain and the brain has processed and responded through speech that the organism is safe.

The posterior system of limbs and digestion also has its own middle system — the reproductive system, which is precisely located between the two posterior limbs and the digestive system. It is also highly rhythmic at the level of lunar menstrual cycles and the physical rhythms of intercourse. In contrast to the male organ, which is limb-like and outwardly constructed the female organ is inwardly located and situated adjacent to the digestive organs, which provide the womb with essential nourishment. There is also a strong connection between the maturation of reproductive organs and speech organs, especially in males whose voices break at the same time as facial and pubic hair grow. Females share the adolescent maturation of reproductive organs and the exploration of sexuality through language and kissing, although the anterior development (facial hair and broken voice) is suppressed. This is consistent with greater anterior development in males (broad shoulders and narrow hips), and greater posterior development in females (slight shoulders and wider hips) — a classic case of compensation.

Schad recognized these anterior-posterior polarities in all mammals. Beginning with the largest and most diverse of the placental mammal groups — the rodents, carnivores and ungulates — he showed how size, anatomical organization, physiology, behaviour and even colour, can all be understood in a holistic way. He also takes into account the three other smaller placental mammal groups: rabbits and their allies, the insectivores and primates and the penungulates (which includes the elephants and sea cows). As shown in Table 6.3, the small rabbits and rodents share a nervous disposition at one end of the spectrum and the large ungulates and penungulates clearly have similar or convergent characteristics. The middle groups, the insectivores and carnivores, also share a number of distinctive characteristics. In simplest terms, Schad

Table 6.3. Schad's holistic categorization (organization) of the main placental mammal groups based on anterior-posterior morphology, organ system and physiological characteristics

Rodents	Carnivores	Ungulates
Hares and rabbits	Insectivores and primates	Penungulates
Sense nerve	Central or rhythmic Lungs and circulation	Limb-metabolic digestive

noticed that these groups are dominated by the anterior, central and posterior physiological systems, that is, the sense-nerve, rhythmic and digestive (or metabolic) limb systems respectively. Domination in this sense means that the systems tend to be *overemphasized* to some degree. Thus the small rodent or rabbit is too sensitive or open to environmental stimuli and the large ungulate too closed or autonomous.

These observations are important because they show a shift in anterior to posterior organ emphasis across the placental mammals as a whole. Even more significant is the fact that the anterior-posterior shift in physiological organ emphasis is perfectly compensated for by a posterior to anterior shift in actual morphology. Thus, the rodents (and rabbits) tend to have a posterior emphasis, sitting up on hind legs that are longer than forelimbs, with long or emphasized tails and short necks, whereas the ungulates have the opposite organization. They have big heads, often with long necks, horns or antlers (which are never seen in the other mammal groups), powerful front limbs, and often with relatively short or inconspicuous tails. The carnivores have a more balanced anterior-posterior anatomy. As noted below, the exceptions and variations are often most instructive and help support this view of mammalian organization.

The compensation between anterior-posterior physiological organ emphasis and posterior-anterior anatomy illustrates the fundamental 'compensation' principle in heterochrony. One cannot overdevelop (grow) one organ or body part without some influence (loss) on the adjacent body part. For example, the ruminant ungulate's stomach is like a 'black hole': it is a highly developed 4–5 chamber digestive system that suppresses tissue growth or assimilation of body mass in the posterior region and so shunts bodily or somatic growth forward into big shoulders, heads, horns, and so on. Conversely, the rodent sensory and nervous system constantly 'burns up' nervous energy and is easily overstimulated (oversensitized), thus restricting physical growth in the anterior portion of the body and displacing it to the posterior.

The fractal nature of these anterior-posterior trends is also seen in the distribution of teeth in rodents, carnivores and ungulates, which are known for their incisors, canines and molars respectively — situated in outward, intermediate and inward positions on the jaw. The incisors are also narrow with a single root, the molars are wide with bifurcating, limb-like roots. Schad notes polarized distribution patterns for colouration, with many small rodents having white undersides and light brown or honey-coloured coats, whereas large ungulates are often very dark in colour. The carnivores, show a rhythmic alternation of light and dark stripes or spots. Although little-known outside Goethean science circles Schad's work has been favourably reviewed,[36] and fruitfully applied to the study of birds and even extinct dinosaurs[37] both of which show anatomical anterior-posterior polarities entirely consistent with the compensating posterior-anterior organization of physiology seen in mammals. Thus, there are dinosaurs such as *Triceratops* with ungulate like horns and molar-like teeth. Colour patterns in bird plumage are incredibly varied, even dazzling, and have traditionally proved very difficult to explain. However, Mark Riegner has shown that they are highly convergent with those identified in mammals by Schad.[38] Indeed the patterns are recursive (or fractal) and understandable across the vertebrate kingdom. Thus, we can begin to recognize that fish, repiles birds and mammals may show similar or resonant patterns such as light undersides and dark backs, spots, longitudinal or transverse stripes and so on. Such convergent patterns give clues to the 'constitutions' of these animals in terms of organ emphasis and behaviour. Until now the 'art' reconstructing the colour of dinosaurs and other extinct animals has been essentially speculative, despite the ubiquity of imaginative art and the fascination the subject holds for youngsters, dinosaur enthusiasts and even lay persons. Schadian observations have the potential to approach such reconstructions more scientifically.[39]

Independent support for Schad's work comes from the studies of the biologist Adolf Portmann, who noted that the mammalian brain has a distinct anterior-posterior organization. He measured the size of the anterior brain (forebrain) or neocortex (which he referred to as the neopallium) relative to the hindbrain.[40] His results showed quite different proportions for small rodents, intermediate-sized carnivores, large ungulates and primates, including humans (see Table 6.4).

All these organizational patterns also allow us to consider the relative balance of openness (outwardness) and closure (inwardness) in

Table 6.4. Adolf Portmann's neopallium index shows the relative size of forebrain (neocortex) in proportion to the hindbrain in different mammal groups. Note the marked increase in primates and humans.

Biological Group	Relative size of forebrain	Mean relative size of forebrain
Small rodents	1.9–7.2	4.8
Carnivores	13.2–23.3	17.7
Ungulates	12.6–32.2	22.3
Monkeys and apes	38.3–53.7	47.0
Humans	170.0	170.0

these groups. The small rodents are too open; they have a small surface area, relative to their body volume, and they have a short gestation period, expelling their young outward at a very immature or 'altricial' stage of development. In contrast, the ungulates have large volume relative to their surface area, and hold the young inwards for a long gestation, thus producing 'precocial' offspring. Again as a 'perfect' compensation, the small rodents make up for their short life span, small size and volume by producing a large number of very similar species (high diversity but low disparity). By contrast, the large, long-lived ungulates are much less diverse in numbers but more disparate (varied) in form. Looked at as a whole, mammalian organization is complexly, but very delicately, balanced or integrated on all levels (size, diversity, disparity, physiology, anatomy, life span, gestation, colour and other characteristics outlined by Schad and others). Mammals begin to look rather like a superorganism system, with two extreme or polar systems mediated by a central system — the carnivores — which quite literally regulate the population of rodents and ungulates, as is well known in ecology.

The alert and biologically-oriented reader will note that there are large rodents and small ungulates. This is true, but rather than contradict the trends we have noted, the pattern fits our recursive organization patterns rather nicely. For example, among the rodents there are mice, squirrels and porcupines, to name just the main groups:

Mice	Squirrel group	Porcupines

Here again we see a similar organizational pattern to that observed in the mammals as a whole. Any group, like the rodents, can be seen as a fractal reiteration of the whole organizational pattern, with its own distinctive characteristics. The mice are diverse and small; the squirrels intermediate in size and numbers; and porcupines species are few. The mice are nervous; the squirrels have aggressive and carnivorous tendencies (including eating birds eggs and nestlings); and the porcupines digest tough plant fibre (cellulose) like ungulates, and have spines made of the same material as an ungulate's horns, except that they are situated posteriorly. The beaver, a very large squirrel relative, also has this ungulate-like digestive capacity and a dark coat.

We find the same type of fractal repetition of organization in the ungulates. Any non-biologist can readily distinguish between the nervous and digestive characteristics of our most familiar ungulates: horses and cattle. Horses are odd-toed ungulates related to the tapirs and rhinos. They have both sets of incisor teeth but no true horns. The rhino horn is specialized outer skin (epidermis), not bone, situated along the mid-line at the end of the snout. By contrast, true horned ungulates from the sheep and cow families (known as bovids), are even-toed (cloven hoofed) and the horns grow from the frontal bones associated with the endoderm (inner skin layer). Some members of this group have lost their upper incisors. Like the hoofs, the horns are paired and diverge or bifurcate away from the mid-line of the animal. Some members of the pig group, especially the warthogs, develop tusks and tubercles (warts) from the middle skin layer, the mesoderm. The tusks are overdeveloped canines. So we see a familiar but modified pattern repeated in the outer, middle and inner emphasis of skin layers, horn, tusk and tooth positions and shapes as we see across the broader grouping of rodents, carnivores and ungulates.

Odd-toed ungulates	Pig group	Horned ungulates

Deer also present an interesting example within the even-toed (cloven hoofed) ungulates. They have antlers, not horns that are replaced annually, thus they are a 'middle' group that manifest a rhythmic alternation between the condition of ungulates with permanent horns and those that have none. Deer also have canine teeth and a few species will eat meat. Some are even spotted or speckled like carnivores.

We could continue this look at the recursive or fractal repetition of morphological patterns at many levels. Any non-biologist can easily distinguish the difference in character between the average cat and dog when it comes to nervousness or digestive habits. Thus the cat prefers live fresh food, especially meat, whereas the dog is less choosy and will eat regurgitated food, including non-meat products, in a behaviour reminiscent of rumination. Moreover, on the delicate subject of the difference in 'refinement' and manners between cats and dogs, we note that the cat prefers a higher vibration, and will sit and purr on one's chest or shoulders, while the dog unashamedly sniffs and nuzzles the groin, genitals and hind quarters of any passerby, not infrequently treating an available limb or rear end as a sex object!

Polarities or convergences in temperament and behaviour are also obvious among the carnivores. Thus, it is not hard to see the rodent or rat-like characteristics of the small stoat, weasel or pine martin, or to see the ungulate like characteristics of the large bear when it is browsing in a meadow of berries. Likewise among the ungulates, the small gazelle is nervous and skittish with an inclination for tender buds and shoots, whereas the large buffalo is placid and content to feed on coarser, low quality vegetation.

Integrating morphology and consciousness structures

Beginning with humans, we looked at the polarity of organ systems from head to toe and saw how patterns of organization repeat, not just at the molecular (sub-organ) level but also at the organ level and the superorganism level. We have seen how the head and jaw appear as microcosms of the whole organism, and how different emphasis among given species actually creates the characteristics that distinguish rodents from carnivores, ungulates or other groups. This applies not just to overall anatomy, but also to detailed morphology of the brain and jaw and to physiology, behaviour and reproductive strategies. We cannot separate morphology and behaviour any more than we can separate mind and body (see Chapter 1). Thus, inherent in the holistic approach is the appreciation of a whole series of correlations between morphology and behaviour at multiple levels. Based on all we have examined thus far we can compile the following chart (Table 6.5) showing the 'characteristics' and polarities of the mammal groups just discussed. We take into account size and anatomy, developmental timing (heterochrony, chronobiology and biological clocks) and behavioural characteristics that we can characterize as mammalian constitutional types.

Wolfgang Schad has also shown that the above-mentioned morphological patterns also relate to consciousness. Thus, at least in humans, the anterior physiological pole, which is particularly sensitive

Table 6.5. *Typical characteristics of rodents, carnivores and ungulates show polarities of organization on multiple morphological and physiological levels.*

Rodents	Carnivores	Ungulates
Small Size	*Intermediate Size*	*Large Size*
Incisor teeth dominate.	Canine teeth dominate.	Molar teeth dominate.
Tendency to light colours.	Tendency to rhythmic colours.	Tendency to dark colours.
Anterior emphasis of sense-nerve physiology.	*Central* system emphasis (muscular athleticism).	*Posterior* metabolic-limb emphasis (stomach power).
Posterior emphasis of anatomy. Relatively small forebrain.	*Central* emphasis of anatomy. Intermediate forebrain.	Anterior emphasis of anatomy. Relatively large forebrain.
Premature/altricial young.	Semi-altricial/semi-precocial young.	Mature/precocial young.
Many species, many offspring, short life span and gestation.	Intermediate numbers, offspring and gestation.	Few species, few offspring, long life span and gestation.

to light and sound, is associated with waking consciousness, the middle system with the subconscious and the posterior system with the unconscious. Thus, we can begin to talk about analogous rodent, carnivore and ungulate consciousness. Behavioural patterns (see Table 6.6) give us clues as to how these groups interact with the world:

Following the work of W.H. Shelden and Carl Jung, one may see parallels between these different mammal characteristics and human constitutional types.[41] We humans are all capable of nervous rodent-like behaviour, aggressive carnivore-like behaviour or placid ungulate-

Table 6.6. Typical behavioural characteristics of rodents, carnivores and ungulates.

Rodents	Carnivores	Ungulates
Easily frightened, flees — easily dies from fright.	May flee or attack. Both aggressive and cautious.	Often unafraid, may not flee. Gives up life with difficulty.
Hides to avoid stimulation; often nocturnal.	Diurnal and nocturnal.	Tends not to hide. Mostly diurnal.
Simple stereotyped behaviour. Difficult to domesticate.	Mix of behaviours: domestic and wild	Complex social behaviour. Easy to domesticate.
Has a hidden nest or home; does not migrate. Has only very local range.	Has a home den, but may wander large distances then return.	Has no home nest or den; wanders and migrates. Large spatial range.
Unemancipated from, or very 'open to' environment. Over-reactive, neurotic.	Mediates rodent-ungulate population	Highly emancipated from, very 'closed to' environment Calm, placid, ruminating.

like behaviour. The extent to which we express such behaviour depends on many factors such as age, gender, constitutional type and environmental situation.

It appears more than mere coincidence that constitutional types vary along a size gradient of small (ectomorphic) to medium (mesomorphic) and large (endomorphic) and that their corresponding organ systems are head (anterior), muscle (middle) and gut (posterior): see Figure 6.8 and chart. This is not to say that an ectomorph is a rodent and an endomorph an ungulate, but it is undeniable that we recognize non-human mammal-like characteristics in our fellow humans. Our language is riddled with references to the rat, cat, fox, bull and bear-like characteristics of our fellow humans. Every aspect of cultural life from heraldry, to cartoons, sports and vehicles is equally riddled with references to animals. It is no coincidence that cars and trucks, from rabbits, to jaguars, mustangs and ram chargers , are in some way like the corresponding animals for which they were named, not merely in shape but in their attributes and characteristics. Bruce Holbrook tells a wonderful story of speaking with a Chinese sage about how western youngsters no longer have the opportunity to undergo initiations such as confronting or hunting predators in the wilds. The sage just smiled and reminded him that there were plenty of predators (like jaguars) taking young lives on the roads every day.[42]

Figure 6.8. Cerebrotonic, musculotonic and viscerotonic constitutional types and their ayurvedic equivalents (Vatta, Pitta and Kapha) correspond to what we know as ectomorphs, mesomorphs and endomorphs.

Ectomorph (small-skinny)	Mesomorph (proportioned)	Endomorph (large-bulky)
Cerebrotonic Jung's rational analytic Computer type	Musculotonic Jung's sensational adapted type	Viscerotonic Jung's natural feeling type

In males, according to Shelden and Schad these particular constitutional types have greater tendencies towards certain diseases:

Schizophrenia	Epilepsy	Manic depression

This approach, although first explored in western psychology in the twentieth century, has deep roots in eastern medicine and philosophy. For example, in Ayurvedic medicine there are three *doshas* (constitutional types): these correspond very closely to the terms used above in western traditions.

VATA	PITTA	KAPHA
Active restless mind, good short-term memory. Intellectually quick but emotionally insecure.	Active, aggressive, even fanatical minds.	Slow careful thinkers. Good long-term memory. Intellectually slow, but emotionally secure.

So far we have only looked at the better-known and more abundant placental mammals (rodents, carnivores and ungulates). But we belong within another group: the primates. The question arises: do we humans, as representatives of a middle group have a blend/mixture of these different consciousness characteristics (structures)? Are they differently expressed according to factors such as individuality, size, physiology, gender, culture, ethnic group, and so on? The answer may be yes, at least to some degree. G.K Chesterton noted that human courage is a very 'contradictory' phenomenon that has 'addled the brains and tangled the definitions of merely rational sages ... It means a strong desire to live taking the form of a readiness to die ...', often in aggressive pursuits like war.[43] There is an uncanny similarity between this human, 'middle group' trait and that described by Schad for the 'rhythmic' carnivores that exhibit alternating aggressive and cautious behaviour, being equally ready to flee or attack.

Hares and rabbits	Insectivores and primates including Humans	Penungulates

We shall examine these questions further in the sections that follow, but, first we should review the main lessons learned from this approach. First, we have learned that all systems are integrated and coherent, and that we can see them as such without rearranging the biological facts of life. Sciences like ecology, genetics and evolution have always suggested that all is connected but they have only shown a few of the patterns that bind organic systems together. They have pointed to very specific connections such as genes or rather vague interconnections such as ecological webs, and until very recently they have lacked any sense of biosphere-wide or even cosmic-scale coherence of the type suggested by the Gaia hypothesis and cosmological but very theoretical and general theories of everything (TOEs). Where the Schadian approach helps advance our understanding is in showing that the interconnections follow patterns or gradients of organization or organic coherence that are recognizable through simple anatomy on a concrete, visible or macroscopic scale. In essence we can see begin to see the wood instead of just the trees. Portmann remarked that modern biology was preoccupied with the invisible (genes and biochemistry) at the expense of the visible. But it is the big pattern of visible organization that is most easily understandable to the non-specialist. Ask any one to describe the differences between the average dog or cat and they will probably give you insightful Schadian-style answers without even knowing it.

Moreover, in this interdisciplinary age, different specializations must be able to connect in a general way, as no specialist is likely to discover the key to big evolutionary mysteries by learning more and more about less and less. Perhaps we need to learn less and less about more and more! As the theme of this book stresses, understanding the dynamics of human consciousness requires a general blend of biology, psychology and scientific philosophy — all seen in the context of ongoing paradigm shifts. Both the reductionist and holistic paradigms have their intellectual and academic strengths and weaknesses. But however we regard them, they represent an inherent polarity characteristic of the mental consciousness structure worldview. The question is to what extent the holistic perspective leads us in the direction of an authentic, and possibly new paradigm of integral consciousness.

In viewing the same biological world from a different perspective, our consciousness of organic reality changes. Schad's genius as a biologist has been to look at the organic world from just such a holistic perspective and show the interconnections which point to different groups of mammals as superorganisms. Without intentionally using the metaphor of complex systems organization, fractals or any theoretical or abstract representation of reality, he has nevertheless shown us that organic organization is entirely consistent with such complex systems paradigms. *In a sense he has shown that nature is the concrete example of what scientists are capable of representing abstractly with complex systems theories.* He shows us that the inherent organization of our own biological group (man and mammals) is as deeply integrated, coherent and elegant as any abstract theory could wish for.

To stress this point, let us take one step further the proposition that the mammals are a superorganism and consider mammals in the context of the vertebrate superorganism and within Earth's biosphere. Let us not forget that all organic systems are also at once open and closed.

The biological literature contains much discussion about the ability of animals to control (or not be controlled by their environment). This addresses their degree of independence, 'autonomy' or emancipation.[44] It is particularly well developed in humans — to the point where we worry that we are too oblivious of the environment, and fearful of destroying something we depend on. Schad, and others, have noted that during evolution the vertebrates have become more emancipated from their environment in a series of easily recognizable steps: that is, we have progressively internalized organs beginning at the anterior (brain) and eventually leading to the posterior reproductive system. (See Table 6.7.) Finally, in humans, even the limbs become internalized as hands and feet adopt new external layers (specialized clothing, tools, vehicles, houses and entire buildings) to act as a buffer and means to manipulate the interface between body and environment.

Table 6.7. The progressive internalization of organs during vertebrate evolution.

Fish	Amphibian	Reptile	Bird	Mammal
Central nervous system	Respiratory system	Fluid system	Thermo-regulation	Reproduction
Brain	Lungs	Heart	Viscera	Uterus

This anterior-posterior trend neatly highlights its reverse or reciprocal trend. Simply put, the most anterior organ (the brain) has had longer to develop. At the posterior end of the scale it is well known that the reproductive system becomes more and more internalized during the evolution of vertebrates, but also in plants.[45] We find a classic example in the mammals, with the outward nature of the monotremes (eggs laid outside body), at one pole, and the inward nature of the uterus in placentals at the other pole. In between we have the marsupials which expel the young in an immature state (like rodents) but then take the young back into the pouch to mature to an advanced stage before weaning. (Thus the marsupial expresses both inward and outward gestures.)

Outward	Rhythmic	Inward
Monotremes	Marsupials	Placentals
egg layers	precocial + pouch	placenta

This view (of inward and outward polarity) is highly convergent with the eastern tradition of Yin and Yang. Yang represents the bold, outward exploring male character and sexual organ (extraverted banners waving in the sun). Yin represents the down-to-earth grounded, nurturer, with an inner womb, and contained rather than released energy. In a strict biological sense the very essence of the continuity and evolution of life embodies this principle. Reproduction does not simply mean the act of sexual reproduction. It also implies the product: reproduction of the next generation. So we can draw a family tree and see the appearance of the next generation from the centre of the two poles (sexes) of the reproductive system. This mysterious origin of new life comes literally (physically) as well as figuratively from the middle of the system — requiring both the outward (open) and the inward (closed) gesture.

Male YANG		Female YIN
	Next Generation	

Clearly both the mammals and the vertebrates have progressively developed their inner organs in such a way as to manifest an organized shift from anterior to posterior. However, the compensation principle argues against defining a simple one-way shift. Balancing the trend towards progressive internalization of posterior organs, we note the steady increase in vertebrate brain size. This trend, known as encephalization, is a major topic of evolutionary discussion, and such measures as an 'encephalization' quotient have been applied to creatures from dinosaurs to humans.[46] The process began more than 500 million years ago, and is now culminating in humans to such a degree that we discuss the 'ballooning' of our brains as a pivotal evolutionary event. Clearly there has been a remarkable internalization process in the strict morphological sense of stuffing more grey matter into the skull, but in another sense this process has been accompanied by an ever greater ability to internalize the outer world cognitively. For example, in the sense suggested by Barfield, the mental consciousness structure has allowed the internalization of thought and the transformation of inspiration into imagination.

We already considered the paradox of simultaneous inner and outer growth in Chapter 1. As the individual grows physically and psychologically within their own skin, so the sphere of outer physical and psychological influence expands. Thus, there is a positive correlation between inner and outer worlds. The biosphere has grown inwardly more and more complex while also expanding to more and more previously uninhabited regions, including, with recent human exploration, other planets. As shown in Table 6.8, *Homo habilis* had at best a crude relationship to global space. The species' geographic range was restricted to Africa, and it had a limited ability to fashion recognizable tools. *Homo erectus,* however, fashioned a recognized stone tool industry (known as Acheulian) that spread throughout the old world. These tools had symmetry and even a design based on the 'Golden Mean' causing some archeologists to speculate on *H. erectus'* aesthetic and mathematical faculties (Chapter 5). Tool size was about one order of magnitude less than the individual's stature. In stark contrast, *Homo sapiens* has a diverse tool kit that, in size, ranges for least 7–8 orders of magnitudes above and below body size. In addition *H. sapiens* has a geographic distribution that encompasses the whole Earth, and the moon. Changes in type of dwelling structures and social organization are also pronounced with obvious spatial parameters.

Table 6.8. Shows the geographic range of three species in genus Homo as a measure of relationship to global space. The size of tools, and complexity of dwelling structures relative to body size as a measure of control over local space.

Species and size	Tool kits, size and shape	Home structures & social organization	Geographic distribution
Homo habilis 1.2 m	Crude (+/- 1 om)	None — unknown?	Africa.
Homo erectus 1.8 m	Hand axe (18 cm +/- 1om)	Caves and shelters. Band-tribe.	Old world.
Homo sapiens 1.8 m	Many (+/- 7 om)	Complex dwellings & complex societies.	Whole Earth and moon.

Thus, expansion or evolution (in the sense of outward development) allows for more inner space (involution). The very human concept of conscious reflection, where outward perception is again reflected inward, draws attention to this paradox of inner and outer experience, and highlights the idea of the outer (open) and inner (closed) polarities of organic systems.

We may not have fully answered the question as to who we are and how we got that way, but perhaps at least we have placed who we are in a broader, mammalian context. Instead of looking merely at humans as a particular species of ape, we can see ourselves as having a more intimate integrated relationship with all mammals and vertebrates. As our consciousness expands both outward and inward, our science and philosophy discover ever broader and deeper evolutionary and cosmic relationships.

It is these notions of yet broader relationships to evolutionary and cosmic processes that we shall examine in the next chapter. They, too, are products of our evolving consciousness.

7. New Evolutionary Paradigms: Coherence, Unity and Beyond

All for one and one for all

We hope by now we have at least made a few points clear. First we emphasized holism and integration. Ostensibly this means that we take as broad a view of consciousness as possible: for example, we include the unconscious, the subconscious and waking consciousness under the consciousness umbrella. These structures, like the five identified by Gebser or the eight discussed by McIntosh, coexist within us as individuals and as a species, though in any given case one or more may be more dominant and easy to recognize, while the others are more latent. Second, as a natural consequence of our holistic inclinations, we attempt to take a pluralistic approach. This means that all consciousness experience can be considered as real and potentially useful in giving insight into how consciousness changes and evolves dynamically. This does not mean that all experience is healthy or typical. In every culture there is a strong tendency towards recognizing pathological behaviours as manifestations of abnormal or unbalanced consciousness, and in most cases we punish, restrain or sanction extreme behaviour if it is considered dangerous to the individual, or antisocial and detrimental to the common good. Paradoxically, at the same time, we tend to revere unusual or abnormal manifestations of consciousness such as genuine scientific and artistic genius, clairvoyance, mystic illumination, spiritual gifts and so on. In such cases we tolerate quite eccentric, even antisocial, behaviour if it is not outright illegal, dangerously extreme or morally reprehensible. Many spiritual leaders fall in the category of having had unusual consciousness experiences. As this may endow such gurus with qualities which attract disciples one may note that, like the reverence offered the gifted mind, the unusual consciousness experience, epiphany or 'revelation' can have a powerful influence over individuals seeking enhanced intellectual and psychological adventure.

Although holism demands a broad perspective it is difficult to consider every *bona fide* sub-discipline within the field of consciousness studies and quite impossible to be aware of the myriad individual hypotheses that emerge on an almost daily basis.[1] For example, we have said nothing of the different functions of the brain sometimes referred to as the 'neural correlates of consciousness'. This is because there are many excellent treatments of the subject, which by definition are brain- and neuroscience-centred.[2] For example, while admitting that the study of consciousness is 'an impossible science', Anthony Freeman shows us that this impossible task is, nevertheless, being tackled from many angles, if not from all the angles we discuss. Freeman cites ultra-reductionist discussions about visual perception which suggest a one-neuron-for-one-purpose approach. Such hypotheses have even led to labelling 'grandmother cells', meaning those cells specifically used to recognize a specific object like grandma.[3] This reminds us of the outmoded one-gene-for-one-purpose view of genetics. In contrast our approach is much broader.

We have by no means ignored the brain, but we have considered it in relation to other major organs of the body, and suggested that it is indeed the centre of gravity of the nervous system and has a special relationship to sensory systems and waking consciousness. We have also shown that we are aware of the increase in brain size as part of a much broader evolutionary trend towards encephalization and anteriorization of primates and indeed all vertebrates. However, we have not discussed the field of evolutionary psychology, a recent spin-off of sociobiology which in turn derives from Darwinism.[4] Debates in these fields speculate on the survival value of all manner of physical, emotional, mental and social faculties and behaviours ranging from language acquisition to rape and altruism. Here we make the provocative suggestion that our preoccupation with survival (or survival of the fittest) is just a scientific manifestation of the existential question of immortality expressed at a purely physical, materialistic level.[5] As the child's questions indicate, the perennial metaphysical conundrums such as, 'Who or what am I?' 'Where do I come from?' and 'What happens after death?' are for many of us seemingly inescapable manifestations of self-consciousness. Science has taken these individual questions and projected them on to the cosmos, biosphere and the human species. So we now ask how the universe, life and our species *Homo sapiens* originated or 'came into being' and how long they are destined to survive. (Paradoxically, although we know that all things must change, anything that is not eternal creates existential angst.) It is rather touching that we transfer selfish concerns about our individual, longevity and wellbeing to concern for the survival of the species, and even other species with whom we share the biosphere. We are even concerned about the fate of the universe. All for one and one for all!

Darwin made many valuable and seminal contributions to our understanding of organic evolution, and the Darwinian doctrine of natural selection certainly explains how poor physical specimens succumb to pressures in the environment that may allow 'fitter' individuals to 'survive' and produce the next generation. This theory of evolution is centred on physical continuity in the process of reproduction and inheritance. Given that survival thus defined depends on the fitter individuals out-competing the less fit, the theory involves tooth and claw selfishness. This obviously explains how we get theories involving *The Selfish Gene*.[6] However, as Darwin himself understood, humans are altruistic and will unselfishly give up their lives to save others. In *The Descent of Man,* he used the word love many times (see further, Chapter 9 below) and even stated that, unlike animals, humans were capable of living by the 'Golden Rule' and were therefore not bound by natural selection.[7] Animals do the same for their offspring, although often only before they are weaned. This central tenet of sociobiological theory raises interesting questions about how much we project the human notion of selfishness or unselfishness on to nature. Regardless of whether such psychological vocabulary is appropriate for other species, it is evident that our collective co-operative consciousness is infiltrating our evolutionary theories to soften the overtly competitive tone.

Thanks to the work of Lynn Margulis it is now well established that organisms cooperate 'symbiotically'.[8] Such cooperation, as in the case of ravens leading wolves to dead elk so they can open their tough hides, or honey rollers leading Africans to bee hives, is not just a day to day convenience, it is also woven into the evolutionary process. For example, it is now accepted that complex cells with a nucleus (known as eukaryotes) evolved early in the Precambrian, some two billion years ago, by the fusion of simpler pre-nuclear cells known as prokaryotes: a process know as 'symbiogenesis'. Interestingly, this notion was initially presented as the invasion of one cell by another. This scenario involves selfish, rape- or intruder-like behaviour depicted in masculine military jargon. In contrast, the cooperative scenario is reminiscent of the mutual benefits of a love affair: all for one and one for all.[9]

We argue that this pluralistic approach is absolutely necessary if we are to arrive at any coherent understanding of the human species and what has been called 'human nature'. As our history, behaviour and creative accomplishments indicate, we are physical, emotional, mental (psychological) and spiritual beings. In common parlance it is an insult to label someone as one-dimensional. Rather we are multifaceted, and so it follows that no single physical (material), emotional, intellectual or spiritual explanation will suffice to adequately describe us. Thus, as our interdisciplinary studies progress, arguments about whether we can explain the body-mind in materialistic, biochemical, electromagnetic or quantum field terms melt away, and we increasingly realize that we are an integration of these multiple phases of energy and matter.

The body electric: energy fields, brainwaves, subtle bodies

Concepts such as the 'electromagnetic genie of consciousness' sound supernatural and exotic, but they are becoming part of the vocabulary of consciousness studies. Likewise, a body described in terms of energy fields or 'information systems' also resonates with early twenty-first century vocabularies of networks and information. But what about gaseous ionic envelopes, molecular coherence, complexity, fractals, recursions and all that? Do they all fit in 'coherently' as part of the big holistic picture? In the following sections we shall examine these various themes and weave a picture from the physical science evidence, considering how physical, chemical and electromagnetic paradigms fit together in the discipline of consciousness studies. We shall then look at some evolutionary theories that provide complementary alternatives to Darwinism without necessarily rejecting it entirely.

We are accustomed to think of our physical body as flesh and bones, although we know, according to the physicists, that it is made up of tiny particles whizzing around in almost empty space. But when we bump into a sharp object we bleed or bruise. The physical body is real. Early studies of human biology began with the dissection of corpses, and the detailed drawing of muscles and organs. Leonardo da Vinci was famous for such pioneering work, which, in his case, combined masterful anatomical study with his unparalleled gift for the fine art of figure drawing. Generations of anatomists and lesser draftsmen followed long before anyone thought of the body in terms of chemistry or electromagnetism.

Since the eighteenth century, biological science has looked beyond our physical bodies to its chemical or biochemical constituents, though long before this the chemistry of alchemy and the dangers of poisons were well known. Any time we take in nutrients, supplements or medications we are interacting with the body as a biochemical system. Indeed, on one level, it is a biochemical system. As modern medicine developed, through the Renaissance and the Enlightenment, we learned to see our bodies as physiological systems in which the circulation of blood and oxygen, respiration, digestion, assimilation and excretion are essential processes. Every physiology student learns about Krebs' cycle, identified by Hans Adolf Krebs in 1937, and otherwise known as the Citric Acid cycle (see Figure 7.1).[10] In this cycle, nutrients (carbohydrates, proteins and fats) are broken down into water, excreted as CO_2 and usable forms of energy. One anecdote tells of a clever student who wrote across the complex Krebs' chart :'Know thyself'. Bearing in mind that the Krebs' cycle knows itself better than we do, and has plans to deconstruct any menu item we might select, biochemistry is obviously essential to the body, and in western society many people are preoccupied, if not obsessed with what chemical substances one should or should not ingest in order to maintain good health.

Figure 7.1. An amusing twist on Krebs' Cycle: "Know thyself." Inspired by an anecdote from Rupert Sheldrake. (Sheldrake 1994.)

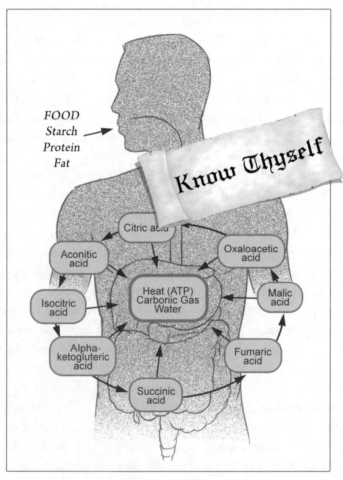

KREBS CYCLE

But nutrition is a tricky business. What natural or synthetic foods, substances and medications (drugs) should one ingest when, and in what quantities? Pharmaceutical companies have grown rich on their biochemical escapades into the world of potent and ostensibly legal drugs. Those who joke about better living through chemistry, include a whole range of caffeine swillers, sugar junkies and those taking antidepressants, tranquilizers, antihistamines and a long list of other questionable products, including those reputed to improve sexual performance. But ingesting these substances not only affects the body's biochemistry, it affects ones mood, psyche and energy level. This is why some substances (many illegal in most societies) are called mind-altering or consciousness-expanding. Yet, as has been increasingly pointed out in recent years, the body constantly manufactures, all manner of mind-altering chemicals, whether under conditions of stress (adrenaline) or relaxation. Indeed, the relationships between biochemical neurotransmitters and various states of consciousness have led to a veritable cascade of literature about the relationship between dopamine, its chemical allies and a sense of wellbeing. Remember that genes, and the genes that switch on, shut off and otherwise regulate other genes, are also all biochemicals.

Before we get caught up in a biochemical or genetic excursion, we should turn to another fundamental attribute of the living body: energy. What is meant by 'energy' either in the simple context of the energy provided by food, or in the far more ambiguous sense used when we talk of our energy level, or even our emotional, or mental energy? Here, we are not talking merely of the citric acid cycle producing biochemical molecules rich in stored energy, for, strictly speaking, energy is not measured in molecules.

We all know when our natural energy level is up and we are not depressed, or artificially stimulated. In many traditions such as yoga and meditation, the body is viewed holistically as an energy field, or as a field of consciousness. Indeed, the main aim of yoga (meaning union) is to achieve 'a perfect union between the self and the infinite'.[11] Thus, with no thought of Krebs' biochemistry, the ancient yogis said: 'Know yourself as the field'. The existence of the 'chakras', meaning 'wheel' or 'vortex of energy' in Sanskrit, has been conventional wisdom in Eastern yogic traditions for centuries: that is, since the time of Buddha around 500–600 BC.[12] In fact the tradition has deeper roots (~1500 BC) in Hinduism and relates to the life energy concept known as *prana* (or *prajna).* Although little known in the West until quite recently, one can now study or learn about the chakras and *prana* energy in almost any yoga class, or holistic health setting. Chakras refer to seven energy centres located along the axis of the nervous system, and referred to as non-physical centres of consciousness. If they are referred to at all in physiological terms it is in connection with the nervous system, and sometimes the endocrine system. As detailed below in Chapter 9, there is also a hierarchy of chakras from lower to higher, and each has its own lower and higher frequency. Here we run into the old 'mental' problem of measurement. If higher consciousness is related to attaining resonance with, or tuning into, higher frequencies it assumes something of a scientific (electromagnetic) connotation rather than the notion of a higher rung on some spiritual ladder. As we have stressed before the spectrum of consciousness involves many frequencies that coexist within us, whether latent (unconscious) or conscious.

So, in short order, our view of 'the body' has undergone several paradigm shifts, from physical flesh and bones, to biochemical vat, to a non-physical vortex of consciousness measured in terms of electromagnetic frequency. This shift in focus was conveniently summarized by the Chinese researcher Charlie Zhang in a paper entitled: 'Electromagnetic versus the Chemical body'.[13] This does not mean that the body is not physical or chemical; it is both, and it is electromagnetic besides. As we shall see, it is possible to describe the body and consciousness in terms of a hierarchy from physical (tangible and low frequency) to psychological or spiritual and intangible (high frequency) states. This, of course, revisits the body-mind discussion. However, our purpose here is to look into how these different paradigms of the human body have been presented in discussions of consciousness, and then to show how these different descriptions integrate and resonate with our themes of unity and coherence.

Many traditions view the human as a coherent complex of the physical and non-physical — what modern gurus like Deepak Chopra call the body-mind. In the anthroposophical tradition of Rudolf Steiner, the human has integrated physical, emotional (etheric), mental (astral) and spiritual bodies (see Table 7.1). A similar scheme was proposed by Willis Harman who contrasts a top down, mind-over-matter metaphysic, giving pre-eminence to spirit, with a bottom up matter-over-mind metaphysic which sees mind or consciousness arising from the material body. G.K Chesterton used exactly the same vocabulary, but adds the twist that the bottom up, inner god arising from matter is 'immanent', whereas the top-down god is 'transcendent'.[14] In either case we can clearly appreciate how one can look at the body of the human being on a multiplicity of levels. These two schemes appear to place the intangible realm of spirit at the top of the table and the physical body at the bottom. But this distinction is dualistic and unnecessary except perhaps as a visual aid.

Historically, western science has tended to work from the 'bottom up', as done in this chapter, first dissecting the physical body, then its biochemical attributes, and later, as the necessary technology was invented, its 'intangible' electromagnetic properties. From here, it is a short step to recognizing the essential mind-body unity and acknowledging the essential reality of psyche and spirit. By contrast, many eastern and western religious traditions, look at the human being from the 'top down' as a physical manifestation of a conscious universe.

But we need not be vague about differences between the empirical, bottom up traditions of science and the metaphysical top down traditions of yogic spirituality and consciousness. Science now provides us with plenty of evidence for the essential unity of the body-mind. Studies of brainwaves for example make it abundantly clear that as we develop through infancy to adulthood, our brainwave frequencies change, and become more and more complex.

In a study by Gordana Vitaliano, eight levels of consciousness are described (see Table 7.2).[15] The first four are very similar to the schemes (above) of Steiner and Harman, but in this case the types of brainwaves (*delta, theta, alpha,* then *beta* and *gamma*) are correlated very precisely with the developmental stages (I–VIII) that humans typically go through during maturation. However, we must stress that most individuals do not develop much beyond stage IV or V unless they devote themselves to consciousness practices. Here again the mental

Table 7.1 A comparison between the schemes of Rudolf Steiner and Willis Harman
in which the human being is perceived as having physical, biological, mental and
spiritual bodies or attributes. The different schemes correspond broadly, but not
exactly.

Anthroposophy Rudolf Steiner	Willis Harman levels	Health Example	Evolution	'Perspective'
Spiritual	Spiritual Science	Spiritual health: wholeness	Universal Purpose	Teleological 'top down' metaphysic
Astral Mental-psychological	Human Science	Individual bio-logical health	Individual purpose	↓
Etheric (emotional biological)	Life Sciences	Organ function: illness	System function; Natural Selection	↑
Physical matter	Physical Science.	Metabolism etc.	Molecular Biology	Reductionist 'bottom up' metaphysic

consciousness inclination to measure levels 'above and beyond' level
V intrudes into our discourse. As description of any level or state of
consciousness depends primarily on experience, we are constrained by
the descriptions available to us from those who report on these states.
We can only really verify an experience directly by having it ourselves.
However, there are also indirect means of verification. These include
recognizing internal consistency in a variety of reports from many
sources. We may also have an intuitive sense that particular individuals
manifest distinct consciousness structures that we recognize as evidence
of a deeper wisdom and more profound grasp of reality, what Bucke
referred to under the label of 'cosmic consciousness' as a heightened

intellectual and moral sensibility (see Chapter 3 above). Recognizing such attributes in another is, in principle, no different from recognizing another's ability as a mathematical or musical genius. What is also of interest here is the desire of the 'seeker' or 'aspirant' to attain other states of consciousness, to develop faculties beyond those of everyday consciousness, perhaps by becoming a disciple of a guru or tradition. In most authentic traditions this is done by long and laborious practice which deliberately shuts out many of the distractions that preoccupy familiar states of consciousness. Despite significant contemporary interest in consciousness, and such fields as psychotherapy, the vast majority of individuals tend to regard their consciousness as 'normal' and as a result do not seek to experience exotic, yogic adventures in consciousness.

As shown in Table 7.2, Vitaliano demonstrates both how our brainwaves change as we grow and how these dynamic changes are linked to physical maturation regions or organs of the brain. He even suggests how these changes may have manifest during the evolution of our hominid ancestors from *Australopithecus* through *Homo erectus* and *Homo sapiens*. Vitaliano's outline of individual development borrows from the concept of the 'triune brain' put forward by Paul MacLean, (see Figure 7.2) and is also entirely consistent with Piaget's childhood development studies.[16] Vitaliano's suggestion that our ancestors may have undergone a similar species level development with parallel shifts in brain, brainwave and consciousness development is intriguing, and similar to the ideas of Jean Gebser. In this regard it is important to note that Vitaliano's higher levels (V-VIII) do not involve any type of ascent to stratospheric 'heights' but are in fact a reintegration with lower levels. Both Ken Wilber and Steve McIntosh have also pointed this out very clearly by suggesting that integral consciousness is a process of ever-greater inclusion.[17] One might characterize this process as an ever more coherent grounding. This reminds us of Gebser's reference to an 'intensification' of consciousness which allows a tuning in to all frequencies at once, or Teilhard de Chardin's ideas on the complexification of consciousness discussed further below (Chapter 9). We are also reminded of Barfield's cycle of primary participation, separation and final participation. Indeed many of the Eastern traditions

Table 7.2. *Correlations between levels of consciousness, human developmental stages,*
brain organ and brainwave development and inferred evolutionary development
(after Gordana Vitaliano, 2000).

CONSCIOUSNESS LEVEL	STATE AGE	BRAIN ORGAN & BRAINWAVE	EVOLUTIONARY STAGE
I Perceptual Body Consciousness	Primary Dualism 0-6 months	Reticular brain stem *delta* 0.5-3.5 Hz	*Australopithecus*
II Emotional Existential	Secondary Dualism 6 months–2 years	Limbic Cerebellum *theta* 3.5–8.0	*H. habilis* — *H. erectus*
III Symbolic Ego Consciousness Pre-operational	Tertiary Dualism 2–7 years	Tertiary Cortex *alpha* 8.0–13.0 Hz	*H. neanderthalensis*
IV Rational Persona consciousness operational	Quaternary Dualism 7–15 years	Inter hemisphere *beta* 13-30 Hz *gamma* 30–50 Hz	Modern *H. sapiens*
V Creative Self-Consciousness Vision logic	Quaternary integration	Inter + intra hemisphere *gamma, beta* and *alpha*	Buddha *manovijnan* Hindu *stulasarir*
VI Supra Individual Symbolic Visions	Tertiary integration	*gamma, beta, alpha* and *theta*	Buddha = *mana* Hindu = *suksmasaria* or subtle body
VII Trans individual Audible illumination	Secondary integration	*Gamma, beta, alpha, theta* and *delta*	Hindu = *karanarsarir* or causal body
VIII Universal Unity consciousness	Primary integration	Frequency = zero	*Nirvana, samadhi, satori*, enlightenment void = form = void

warn against the dangers of detachment as a result of intense 'higher' consciousness experiences. So, the accumulating evidence seems to suggest that forgotten Hindu and Buddhist pioneers understood and experienced altered, higher states of consciousness through deliberate practice, long before they were measured and experienced and reported *de novo* by western researchers and given new scientific, electromagnetic field labels. Nevertheless, as Vitaliano demonstrates, the western mind has not been entirely oblivious to these insights. Moreover, Piaget, Gebser, Barfield, Wilber and others have pioneered an understanding of the dynamic flux of evolving consciousness and explained it as an essential facet of human experience and existence.

Figure 7.2. The triune brain. (After Paul MacLean 1990.)

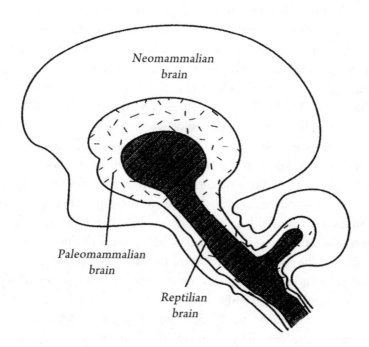

Enveloped by consciousness: gaseous genies and Russian dolls

Vitaliano points out that it is possible to measure a number of biophysical properties of the human body in the 'extremely low frequency' (ELF) range. These can be termed subtle energies (and how charming and coincidental that they are associated with the symbolism of elves). Unlike the rather slow biochemical information processing that goes on in the brain a much faster biophysical processing goes on in a 'gaseous ionic optical system' or 'envelope' located outside the body. This is a manifestation of the human energy field, and might also be termed an 'aura' visible to some but known to others through measurement with delicate instruments. Such scientific revelations raise fascinating questions about where the body begins and ends. The evidence suggests that this gaseous envelope is in a constant state of flux, which in turn is related to states of consciousness. Vitaliano calls this gaseous envelope, the 'genie' that hovers outside the vessel of the body. It is interesting that the envelope is most diffuse and developed in infants during the primary body-consciousness stage, and that it begins to contract and shrink into conformity with the contours of the physical body as development proceeds through the next stages (II through IV). It is almost as if a diffuse, universal consciousness is absorbed by the body as it matures. The result is that an external consciousness becomes concentrated in the body and crystallized as self-consciousness. This is also figuratively and linguistically what happens as we 'take in', or become conscious of, the world around us. In our opinion, this view of consciousness is remarkably convergent with Plato's notion of an ideal intelligence or mind that descends or incarnates in the individual in an inferior, less-than-ideal form. This is the top down metaphysic of Harman (see Table 7.1), and an alternate view to the oft-cited puzzle of materialist or bottom up advocates who wonder 'how consciousness arises from matter'. Symbolically at least, one might regard this internalization of consciousness as the entering of the soul into the body. Here it pays to recall that evolution has involved repeated phases of internalization (see Table 6.7). Clearly the process is coincident with the emergence of self-consciousness and language (stages II–III). This concentration or 'solidification' of consciousness in the rational stage of development, actually blocks many of the subconscious, unconscious or psychic attributes of simple consciousness, and it allows us to forget that consciousness may, in some sense, have been 'delivered' from the top down. Thus, it is only through continued or expanded development of consciousness through stages V–VIII that higher and progressively more universal consciousness can penetrate back down through the buried layers to integrate our rational, symbolic, emotional and perceptual consciousness. In these stages the genie again escapes from the bottle as consciousness expands beyond the body's boundaries. Indeed Wilber's *No boundary* metaphor is apt for the journey toward universal consciousness.[18]

Vitaliano also uses the metaphor of the Russian dolls for the build up of consciousness through stages I–IV. At each stage an earlier diffuse consciousness is confined, encapsulated, constrained, suppressed or over-ridden by the power of the growing self. Then in stages V–VIII these captive consciousness structures, now held in doll shells of the buried subconscious and unconscious must be liberated and reintegrated in the full light of an expanded, higher consciousness — in the 'transparent' light of Gebser's intensified, integral consciousness. Each closed doll must be opened down to its primary core. When the innermost doll is opened there is only emptiness: the unconstrained, unconfined, boundless universal void, where no defined structure imposes.

William Irwin Thompson described this conscious reintegrating 'down' through the first four levels of consciousness as similar to what the Buddha had done when he achieved enlightenment in the Axial Age around 500 BC . The difference between the chart we have compiled from Vitaliano's work and that of Steiner and Harman, is that it avoids the unintended visual dualism of 'bottom up' versus 'top down' schemes. Wilber has show that the similar schemes can be presented both as linear, ladder-like scales or as cycles. (See Figure 7.3.) In these schemes we have an expansion-contraction cycle, resonant with the dynamic flux inherent in organic systems, whether it be in the inhalation and exhalation of respiration or the larger cycle of birth and death. The eight stages bring us back to where we started, but with the difference that we have evolved from a simple unconscious state (Id) through the Ego state to one of fuller universal (Superego) consciousness. This is the potential of what the adepts call the consciousness adventure.

Figure 7.3. Levels of consciousness as a reintegration cycle. (After Wilber 1986, Fig. 37.)

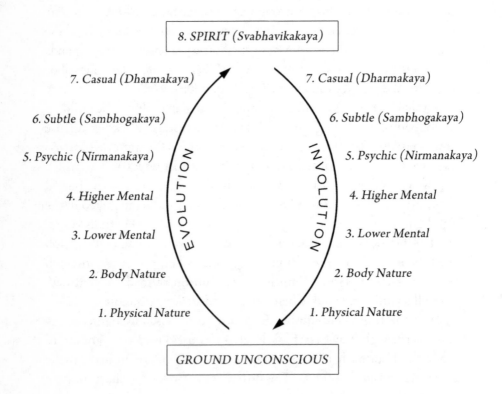

Barry Long put it equally well when he said that early, primitive humans were highly, but unselfconsciously psychic before they came into their bodies. But, when waking consciousness emerged it suppressed unconscious psychism with a flood of conscious sense perceptions with which individuals began to identify the self — thus creating a discrete centre of consciousness. So as the universal connection was lost, the individual fell out of the timeless spiritual world into time, with all the pitfalls this entailed, including fear of death and the angst that arose with Barfield's 'separation' from our universal, ever-present origin. In Long's opinion it is only in consciously regaining our psychic powers that we advance in the process of reintegration.

If any of this has a familiar ring it might because we respond intuitively when we know something to be or feel 'right' — when something 'already known' on an unconscious level comes into consciousness. T.S. Eliot put it succinctly in his famous 1922 poem *The Waste Land:*

> We shall not cease from exploration
> And the end of all our exploring
> Will be to arrive where we started
> And know the place for the first time.

Molecular democracies and spooky quantum quandaries

Mae-Wan Ho, the British geneticist turned biophysicist, described the body as a coherent molecular democracy in which every cell arranges itself with its optical axis in alignment with the body's axis. It turns out that our tissue, our mere flesh and bones, has remarkable properties. Far from being dense physical matter, our 'wet' organic systems are actually rather like large liquid crystals (with gaseous ionic envelopes). Each cell has its reservoir of stored energy (as Hans Krebs would no doubt be pleased to learn.) Given the right measuring devices, this energy can be seen as light or biophotons. We are literally vessels of light — figuratively the 'Inner Light' of Chesterton's immanent God, noted previously. The complex world of biophoton research, best penetrated with a higher degree in optics, is now a major field of research. Ostensibly these are the subtle energies known to the yogis and described in their ancient consciousness traditions. The light and energy spectra of the chakras should be no surprise, as light is ubiquitous in the universe and inseparable from the fabric of the electromagnetic spectrum.

While Ho's research involves complex physics, it also has profound implication for consciousness studies. Her work on 'The biology of free will', published in the *Journal of Consciousness Studies,* suggests that the stored energy of cells and organisms makes each an autonomous entity able to act independently, freeing itself from certain laws of physics: namely, the Second Law of Thermodynamics.[19] Ho's conclusions are convergent with those of Zhang, Vitaliano and the ancient Hindus and Buddhists, all of whom hold that the human organism can be viewed as an energy field. As described earlier, it is well known in these ancient eastern traditions that the concentrated energy of the nervous system — known as the kundalini (or coiled serpent) — can rise from the lowest to the higher or highest chakras, often producing an instant sense of illumination and life-transforming shifts in consciousness. Although not an everyday experience for most people, this kundalini release or 'rise' is usually considered as a spiritual awakening (see here Chapters 3 and 9) or enlightenment that is almost always discussed as an activation of evolutionary energy. In terms of Vitaliano's scheme, a kundalini experience is a journey from the constrained domain of individual Ego/persona consciousness to one of greater universal spirituality. Such experiences could be referred to as a type of democratic mass movement, as the stored energy of the whole body system mobilizes in coordinated fashion.[20] Certainly the metaphor of an energetic 'phase shift' is also appropriate. It also appears more than mere coincidence that the flow of energy is from the lower (posterior) to the higher (anterior) organs, although not always reaching the highest chakras. This direction of energy flow is essentially the same as that outlined by Piaget for child development or by Gebser for long-term human evolution. The main difference is that the timing is greatly accelerated. As we shall see in the next sections, such differences in developmental timing between humans and other species probably play important roles in evolution.

Although the modern scientific view of the human body-mind has tended to dispense with any notions of a real spiritual dimension by substituting a material-physical paradigm, twentieth century science has led us inexorably to new biochemical and electromagnetic field paradigms, while psychology and consciousness studies have also developed on a parallel tracks. The resulting convergence in the field of subtle energies has pointed squarely to the insights of ancient wisdom traditions, allowing us to see (and scientifically verify) deeper levels of reality, previously penetrated or intuited only by the mystical minority. All this might suggest that science and knowledge in general is in some way related to a primary datum of universal truth waiting to be discovered 'out there', like Plato's universal mind. Creative level (Vitaliano's level V or higher) insights emerge from the minds of scientist-thinkers because our/their consciousness is dynamic and evolving, allowing us to propose new evolutionary paradigms and propose new ways of looking at reality. However, inasmuch as emergent consciousness, has to come from somewhere — figuratively speaking — the evolving brain is not so much producing new consciousness as it is tuning into the infinite universal source. Scientists address this when yearning to known 'God's thoughts', and in this regard their quest is essentially identical to that of the mystic.

The convergence of quantum mechanics and consciousness studies provides an excellent illustration of this ability to formulate new paradigms as a result of tuning into subtle energies. Those, like Roger Penrose, who claimed rather cryptically that consciousness 'is connected with quantum gravity ... exact nature unknown', are naturally predisposed to consider quantum mechanics because they are among a relatively small mathematical élite capable of understanding a field whose intricacies are a mystery to most of us.[21] Quantum mechanics, like consciousness has potentially universal applications, and both deal with the ultimate in subtle energies. The best example is probably the idea that reality is 'nonlocal'. This means there is mysterious or 'spooky action at a distance' (*spukhafte Fernwirkungen*) in which the act of observing minute particles (quanta) or waves not only affects or 'collapses' their properties but also affects (collapses), the properties of other particles.[22] At the risk of gross simplification this means that discrete objects are no longer separate. In a quantum universe, it is as if they are telepathically connected. So, just as it was once acceptable to talk about a mechanical universe, where the mass of one planet affected that of another through gravitational attraction, we can now talk about a quantum mechanical universe, a conscious or intelligent universe or even a conscious quantum universe. Quantum thinking allows for very satisfying, even awe-inspiring, insights that lead to mind-body unifying statements like: 'being physical no longer means being material but being a structure in space and time that somehow holds ("encodes") knowledge or information created by earlier events'.[23] This almost seems to paraphrase pop psychology's 'the issues are in the tissues' — or perhaps more precisely, 'the issues are the tissues'. Such ruminations also remind us that as we analyze what it means to be a physical entity, the reductionist approach of dissecting bodies into tissues, cells, genes, molecules and subatomic particles ultimately leads us to the purely intangible realm of non-material ideas. 'A structure in space that somehow holds ("encodes") knowledge' serves as quite a passable definition of consciousness.

We should not mention Roger Penrose and his physicist colleagues without also briefly discussing the Anthropic Principle, which in simplest form suggest that the laws of nature are so perfectly balanced that without them the universe, life, humans (and anthropology) could not exist.[24] Penrose asks: 'Are the laws of physics specially designed to allow the existence of conscious life?' Presumably there are those of various religious persuasions who would simply answer yes: a God or designer created things 'just so' for a good reason. Indeed a common tenet of nineteenth century natural theology and natural philosophy was that the world had indeed been created 'just so' for our benefit. Here, as one might imagine, we run into a significant controversy in science. Those who see the universe as purposeless, the result of random processes, albeit determined by impersonal 'laws' such as entropy, which leads to dissipation of energy, and ultimately universal decay, consider it wishful human thinking to consider that the universe has a purpose, goal or objective.[25] The problem of cosmic purpose known as teleology, has become a perennial focus of debate in scientific and philosophical circles.[26] Although rejected by many scientists as speculation, it is physicists, not scientifically-naive or religious optimists who are most responsible for formulating the Anthropic Principle. This suggests that the universe manifests sufficient order to always raise questions about intelligent designers, or intelligent design mechanisms. Remember that Einstein's reason for proclaiming that he wanted to known 'God's thoughts' was that the more his scientifically-refined consciousness learned to understand the physics of the universe the more he was forced to marvel at the manifest intelligence. He expressed the same sentiments as Ralph Waldo Emerson when he said that: 'we lie in the lap of immense intelligence'.[27] We shall continue this debate as we proceed to explore new evolutionary paradigms and what they tell us about order and purpose in the organic world. As a footnote to this section, and without wishing to insinuate that sceptics are absolutely wrong to see the universe as meaningless or purposeless, it is worth noting the sceptics' dilemma in this regard. If the universe really is meaningless, then opinions of sceptics are also meaningless and can be disregarded! The more positive reciprocal of this argument implies that by engaging in any discussion one has already assumed that the dialogue must have some measure of meaning.

May you be forever young

This happy wish from the pen of Bob Dylan reflects both our fear of aging and death and our desire for immortality. (Don't worry, much evidence suggests we are immortal!) We have already touched on the psychological issues that underpin our preoccupation with survival and our ruminations about the purpose of evolution and how it may affect our survival as a species. This problem of teleology remains particularly controversial in evolutionary studies. Is man (humankind) the pinnacle of creation as suggested by Aristotle's Great Chain of Being, or, as religious vocabulary might have it, are humans created in God's image? Does evolution have direction and purpose? As the cosmos, biosphere and its species and individuals evolve (change) at least for some entities there is a process of complexification, which like the development of a human or higher mammal from a single cell, follows a plan and achieves a complex culmination or objective. The seed's destiny is to become a plant, then a flower, and then perhaps at some point to make way for a new species incorporating some of its ancestral characteristics. We shall revisit this question of directed or straight line evolution (orthogenesis) in our final chapter.

It is common knowledge that humans are primates and that chimps, gorillas and orangutans are our closest living relatives. Not only do we look like apes, but Darwin showed that we behave and express our emotions in similar ways. Thus, it is often said, with equal frequency that we 'descended', or that we 'arose' from apes. But, which is it? Are we ape offspring, or ape grandparents? Are we the result of bottom up, or top down, processes, or both? Such questions fuel many evolution debates. Actually it is ancestral species like *Homo erectus* and the Neanderthals, not the apes, that anthropologists regard as our closest relatives. But because these species are extinct, apes are regarded as our closest 'living' relatives. However, the apes have rather a poor fossil record, and there is much about the origins of these familiar relatives that remains unknown. For these reasons we need to turn to developmental studies for information that helps give us evolutionary insight.

The Belgian chemist, Jos Verhulst, former Waldorf School biology teacher and writer on nuclear physics, has done precisely this. In his seminal book *Developmental Dynamics in Humans and other Primates,* Verhulst examines the important role played by timing or heterochrony (see Chapter 6) in differentiating human and ape development.[28] For example, when it is a seven month old fetus, the chimp has the same hairline as the human, but by the time it is born it is at the same stage of development as a twelve month old human. This indicates that it develops faster than the human in these areas. Thus, the question becomes how does this make a difference, and is it an advantage or not?

Verhulst synthesizes his detailed knowledge of primate anatomy development into a novel theory of evolution which he names 'synergistic composition'. As the label implies, the combination of two or more phenomena lead to the creation of a new entity (whole) that is more than the sum of the parts. Verhulst shows that most human organs, from limbs and hands to heart and lungs, demonstrate what is known as 'retardation' (meaning protracted or extended life span) relative to other species such as our rapidly developing chimp. Even at the 'molecular level' primate and human evolution appears to be 'retarded in comparison to that of the average mammal'. These growth dynamics express as 'early' Type I retardation-fetalization characteristics, or as 'late' Type II development characteristics. So, we can identify new developments at both ends of the life cycle. It is this combination of early (Type I) and late (Type II) phases that Verhulst calls 'synergistic composition'.

This synergy is well known with respect to human reproduction. We reach sexual maturity slowly, have relatively low reproductive rates and then have a long phase of post-reproductive life.[29] Humans are sensitive to various sidereal rhythms such as the lunar and seven year cycles.[30] (See Figure 6.4.) By contrast, ape physiology does not respond to the lunar cycle so we cannot have inherited this trait from them, by natural selection. This suggests an emergent new human trait, and could suggest that human evolution is not random, but proceeds towards some goal of maturation or completion.

The descent of the larynx is a Type I retardation phenomenon that opens the potential for speech — a Type II development. This is a type of time integration where early developmental trends appear as precursors or 'prefigurations' that anticipate latter developments which complete them and give them meaning (rather like the meeting of two complementary creative spirits!) So, unlike the chimpanzee which grows hair before leaving the womb, thus moving quickly in the chimp/animal direction, human 'retardation' (fetalization) allows new potentials to emerge without imposing animal specializations. We only get a little more hair later in life — particularly in males because they grow larger on average. This Peter Pan syndrome leads to longer life and is the biological equivalent of not going 'off track' too early. Another Type I example is upright bipedal posture, which is foreshadowed or 'prefigured' in many animal species during embryonic development, but lost by an early commitment to becoming quadrupedal. Type II developments also include the extended development of the pre-frontal cortex. All this suggests further intriguing conclusions about human evolution. In Verhulst's words 'what remains unspecialized and universal in animals ... this universal, unspecialized kernel, which is the true inner architect of the human organism, reveals itself as the bearer of extraordinary potentials' such as uprightness, speech and culture.[31]

The phenomenon of prefiguration (sometimes called pre-adaptation) is intriguing and controversial to evolutionists because of its teleological implications. Verhulst cites the birth canal of chimps and other large anthropoid apes as much larger than necessary for the skull of their newborns. Thus, unless this feature is mere coincidence, anthropoid pelvic structure appears to anticipate the later evolution of the large human skull. This would suggest that evolution appears to have teleological foresight and implications for future evolution. Thus Verhulst draws some provocative conclusions. 'Enhancement of and liberation from the animal element has not yet peaked in modern human beings'.[32]

This liberation from animal characteristics, does not necessarily imply human superiority. From a biological viewpoint it is just a statement of differences resulting from heterochrony: that is, necessary differences in timing of development which give rise to different species. For example, juvenile *Australopithecus* resembles adult *Homo erectus*, juvenile *H. erectus* resembles adult Neanderthals, and juvenile Neanderthals resemble adult modern *H. sapiens*. (See Figure 7.4.) Thus, infant *Homo sapiens* may resemble a future childlike iteration of hominid evolution — an innocent 'completely liberated angel-like race of beings. This thought touches on the central mystery of human existence'.[33] In this sense, as our day to day language confirms, our infancy, and our children, are our evolutionary future. It may seem counterintuitive to suggest that as evolution progresses, each generation gets younger, but, quite literally, each generation, while having a longer evolutionary history is also at the same time younger than the previous generation. This is the ideal way to avoid becoming overspecialized. An analogy in our social life can be seen in the trend towards increased play and leisure in recent history.[34] Moreover, this way of thinking seems remarkably resonant with the reintegration dynamics discussed in relation to consciousness. We can always return to the 'ever-present' wellspring of youth for evolutionary revitalization: it is no coincidence that this is a recurrent theme in literature and culture. If this is in fact the essential dynamic, then we are in fact, as much the ancestors (juvenile precursors) of the apes and our pre-human relatives, as we are their latter day descendants. (This perhaps helps explains the semantic confusion.) Evolution held back (retarded) our development while in the apes and other hominid species, it was allowed to move partially, but not fully in 'our' direction. But we must remember,

Figure 7.4. *Progressive juvenilization of hominid species: a, b — juvenile and adult Australopithecus; c, d — juvenile (specimen incomplete) and adult Homo erectus; e, f — juvenile and adult Neanderthal; g, h — juvenile and adult Homo sapiens. (After Verhulst 2003.)*

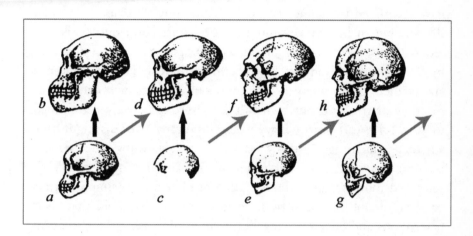

that evolution 'involves' us all in a co-operative effort that transcends time, and certainly transcends our lifetimes. So, in this scenario, we become a less-than-fully-developed stage or 'experiment' that will allow the emergence of a future species that has bypassed or transcended our limitations. (See Chapter 9). Here then is a biological basis for optimism about tomorrow and our inclination to think teleologically about future goals, objectives and purpose.

Thus, Verhulst's conclusions have far reaching implications, and he asks whether 'the materialistic image of the human being, or materialism as a whole, is really true, and does Darwinism adequately explain human evolution?' Verhulst is convinced that it does not. Following the work of Louis Bolk (1866–1930), Verhulst recognizes an 'intrinsic evolutionary factor' that guides evolution towards goals (including the human species) in much the same way as intrinsic factors determine that the seed will become a flower.[35]

Such a goal-directed (or teleological) view may not sit well with Darwinians enamoured of the role played by pure chance, but it flies in the face of biological evidence to deny intrinsic processes that produce such complex 'final' results.

Verhulst's critique of Darwinism emphasizes the failure of materialist ideology to account for consciousness. Philosophers like Daniel Dennett have attempted to characterize or 'explain away' consciousness as an illusion, but such a stance cannot be valid since consciousness is first required to suggest that it is an illusion! Despite the work of quantum mechanics experts, physics and chemistry have not yet produced a law predicting the emergence of consciousness and such concepts as meaning or truth. 'If our own thinking were completely determined by such [physical] processes we would be incapable of distinguishing whether something is "really" true, or merely seems true because physical processes in our nervous system create that impression'.[36] As suggested above, sceptics who actually argue like this are simply victims of severe doubt, having little faith in free will: they hold that their perceptions and beliefs are invalid, without dismissing their sceptical positions as equally dubious.

Verhulst's position is that: 'Life is not some complicated physiochemical process but a phenomenon in its own right'. It is 'not a property of matter; it is a property of form' without physical 'building blocks' *per se*. Life's attributes are unique and transitory and characterized by 'perpetual change ... the continuous alteration of form'.[37]

Verhulst is finally driven to a severe conclusion when he characterizes certain Darwinian representations of evolution as 'mass indoctrination'. His view is that 'life and consciousness are subject to their own dynamic principles'.[38] As he shows, formative processes lead to the unfolding of the organism, and also guide evolution as a whole. They are present as intrinsic factors from the outset. Just as evolution fulfils its potential to raise the flower from a seed, so it also embodies a promise to fulfil its potential in humans and the biosphere as a whole. This does not mean we know what future fulfilment will look like, but it is patently obvious that we attempt to bring about the outcomes we desire. Again good old teleology is our ever present psychological companion!

Table 7. 3 Modes of thinking and their relationship to time and scientific subject matter. Modified after Schad 1977.

Mode of thinking	Temporal relationship	Appropriate subject
Causal-deterministic	Influence of past	Inorganic matter
Instantaneous	Present	Biological systems
Teleological	Pull of the future	Psychology — mind

Wolfgang Schad points out that our thinking is essentially teleological.[39] Almost anything we do involves projecting our thoughts, goals and objectives into the future. This is the essential mode of psychological thinking. By comparison, causal or deterministic thinking is appropriate to the analysis of physical matter and mechanistic problems where past conditions do in fact influence the present and future. (See Table 7.3.) Interestingly, in Schad's philosophy the physical past and psychological future meet in us in the biological present where most physiological reactions are instantaneous or near instantaneous: that is, bound to the present. Thus, we begin to see how we might equally well adopt an incomplete bottom up or top down metaphysic. We should not project one mode of thinking on to the entire cosmos. Teleological thinking is bound to crop up in many scientific and philosophical discussions about the organization of the cosmos, biosphere and the role of humans therein. To deny this is merely to reveal a particular bias. As the philosopher Alfred North Whitehead noted: 'Scientists who spend their life with the purpose of proving that it is purposeless constitute an interesting subject of study'.[40]

Verhulst's work indicates that evolution is driven as much by intrinsic internal dynamics as by the external forces of selection. The late Stephen J. Gould in his final magnum opus — The Structure of Evolutionary Theory — acknowledged the importance of the intrinsic or 'formal' dynamic, an idea that originates with Goethe, as equal in importance to the external 'functional' dynamic associated with Darwin's theory based on natural selection.[41] Simply put, formalism allows the organism to make an intelligent or sensitive response to changing environments, instead of making the organism a passive victim of natural selection. The organism is not a machine or automaton as some archaic ideas hold.[42] Nevertheless, biomechanics is a big field in biology and the rather deficient machine metaphor is still widespread.

Lee Spetner in his book *Not by Chance* parallels this somewhat Goethean view by noting that the success of plants and animals is not controlled only by the extrinsic forces of natural selection. Using detailed genetic arguments, he challenges the claims of neo-Darwinian Theory 'that (*random*) mutations are not at all related to the needs of the organism'. Instead he points out that mutations result from the intrinsic needs and responses of the organism to its environment. Spetner postulates a 'Non-Random Evolutionary Hypothesis' (NREH) which holds that changes in the individual (known as the phenotype) come about 'more like the purposeful throwing of a switch than a random mutation'.[43] Organisms are intelligent sentient beings with the type of inherent, energetic free will discussed by Mae-Wan Ho. There are many fascinating examples. Plants produce more seeds when they are set further apart, rather than close together. Snails somehow know if crabs are around and so grow thicker protective shells, and barnacles have a similar response around predatory snails. This is not the same as the weeding out of thin-shelled forms by natural selection. The snails deliberately 'select' thicker shells for themselves ahead of time. This may be an unconscious response, but it is nevertheless an active response. Moreover, the physiological response to stimulus from the environment (other species) is biologically instantaneous, as Schad indicates (see Table 7.3). They are one intelligent step ahead of the supposed natural selection mechanism. They are acting or 'mutating' proactively rather than waiting passively to become helpless victims.

In another departure from classic Darwinism, Spetner also notes that the evidence indicates that evolution takes place, to a significant degree, at the level of the organism not the population. Large scale transposable genetic elements known as insertion sequences (IS) and transposons bring about major genetic (adaptive) changes in response to environmental signals or stress. Familiar bacteria such as *E. coli* and

Salmonella, if tampered with to remove the genes that perform certain functions, quickly adapt by switching on substitute or cryptic genes to perform those same functions. Likewise, Kurt Goldstein noted that the brain is capable of similar adjustments when damaged.[44] Effectively large scale mutations, supposedly requiring millions of years to accumulate through random neo-Darwinian processes, take place, non-randomly, in mere days. Moreover, these intrinsic biologically driven mutations appear to be inheritable. This theory of inheritance of characteristics acquired during an organism's lifetime is known as Lamarckism, and was for a long time considered antithetical to Darwinism and even to modern genetically based neo-Darwinism. As the type of work cited by Spetner and others indicates, this has now changed, and it is clear that organisms respond actively and intelligently to the environment and can pass on the genetic changes they acquired in short order.[45] All this raises new questions about the intrinsic intelligence of organic systems and cosmic organization, and will no doubt contribute to lively discussion by those involved in the intelligent design debates.[46]

A new evolutionary grandeur

Famously, in reference to his evolutionary theory, Darwin stated, in the final paragraph of the *Origin of Species,* that 'there is a grandeur in this view of life'. He referred, of course, to the pleasing coherence and consistency one perceived when linking all life together in a web of evolutionary relationships. Since Darwin's day many branches of science have postulated unifying theories, that not only demonstrate deeper relationships in the organic world (genetics for example) but also attempt to show deep organizational structure throughout the cosmos.

As noted in Chapter 2, Laszlo's *Evolution: the Grand Synthesis* suggests ways to integrate our concepts of organization in the physical world — the cosmos of galaxies, stars and planets — the biological world of organisms, and the complex domain of psychology and socio-cultural systems.[47] (See Figure 2.3.). In a more recent book *Science and the Akashic Field,* Laszlo launches into a much deeper synthesis.[48] He suggests that the 'quantum (Q) vacuum' and Zero Point field concepts have replaced the idea that space is empty. This Q vacuum is a super-dense cosmic medium carrying density pressure waves that probably set values on gravity, electromagnetic and the nuclear forces. Matter is a secondary wave form disturbance in a sea of energy. We may compare this interesting suggestions with the aforementioned idea that life is not a property of matter, but more of a field, or form-like phenomenon. 'All processes have an inner rhythm according to their resonance with the vacuum's standing waves'. In a truly frictionless medium: *'The quantum vacuum generates the holographic field that is the memory of the universe'.*[49] [Laszlo's emphasis] The universe is thus a unified field — which Laszlo labels the Akashic field, derived from the Sanskrit word *akasha* (meaning 'sky', 'space' or 'ether'). This gives rise to the concept that all Nature's knowledge and information is stored in the ether: that is, on a non-physical plane of existence. This ether does not refer to the discredited nineteenth century concept of a medium to carry light.[50] Given the thermodynamic tenet that matter cannot be created or destroyed, this claim is not contradicted by science. Thus, Laszlo has short-circuited our voyage back in time by linking cutting-edge physics with ancient Hindu tradition, and the traditions of mysticism which we shall examine in Chapters 8 and 9.

This seems to bring us back to the notion of the physical organism as 'structure in space and time that somehow holds ("encodes") knowledge or information'. Ostensibly the universe is highly complex with many layers of interaction such as the interpenetrating timing regimes that organize all the physiological process in our bodies (Chapter 6) and in ecosystems, reefs and Gaia-scale biosphere systems. In such systems that involve the creations of culture, including Laszlo's innumerable books, the connections and 'ties that bind', the modes of communication and interaction, whether obvious or subtle, are so complex as to defy classification in any sort of simple scheme. Thus, it is no coincidence that our consciousness has recently spawned the science of complexity.

So we come to the ultimate level of complexity that we can self-referentially contemplate: the human mind and all the complex cultural institutions which this same consciousness has created, *including the study of consciousness!* These 'sociocultural systems', as Laszlo calls them, appear to be emergent properties of complex biological systems. Just as multi-cellular organisms evolved from simple, single cells, and complex reefs evolved from simple ecologies, so anthropologists, historians and sociologists track and define complexification trends in human cultural history. As examined in Chapter 5, these shifts create complex networks of communication, which manifest in trade, economics, language and media as well as in ethical, moral legal, religious and spiritual traditions. Communication networks, although still involving physical exchange of written documents, goods and other currencies, also increasingly involve communicating many intangibles, such as ideas, cooperation, ethical and moral conventions. These exchanges mostly lack any definable physical manifestations or forces, other than intangibles such as 'market forces' or 'confidence'. Such complex systems of organization are not so much local (atomic, molecular, cellular or individual) as global. They have been summarized, not physically or biologically but psychologically, as the 'noosphere' which carries the connotation of 'sphere of the mind', noetic polity or even a 'global brain'.[51] (See also Table 5.3.) Here we have another convergence of consciousness vocabulary: the noosphere, which has become so evident in the twentieth century as a planetization or globalization phenomenon, is nothing if not a manifestation of a collective consciousness. And how is it different from the mystic's experiential report of a universal consciousness? Let us explore this question in our final chapters.

8. Into the Mystic

Scientific prophecy and global mind change

Bucolic wisdom, Buddhist philosophy and common experience tell us that all things change, often quite unexpectedly. 'You cannot step twice in the same river.'[1] Life reminds us of the changes brought by each new generation, and politicians will inevitably promise change with every new election cycle. Paradoxically then, the only constant is change. But this view is consistent with at least one half of the previously cited philosophy that life is a dynamic tension between conservative and novel forces. However much we cling to the past and however comfortable we are with the status quo at any given time, sooner or later something will change.

In this our penultimate chapter, it is hardly necessary to repeat the evolutionary, prehistoric and historic evidence for multiple timescales of change given in ample detail in previous chapters. So here, in anticipation of a grand, 'integral' finale, we shall attempt to weave together the threads of our physical, emotional, mental and spiritual evolution on a more fundamental level and try to shed light on some of the most basic existential and metaphysical questions: who are we, where did we come from and where are we going? A tall order one might say given that such questions have been posed for millennia, without producing any absolute consensus. But by our very nature, both as children and adults, many of us are compelled to ask these 'spiritual' questions for as long as we live. We acknowledge that many aspects of existence are mysterious and likely to remain so, and that the mystery may deepen the more we learn. Those who do not ask metaphysical questions, or certain types of metaphysical questions, may have good reasons to judge some philosophical and existential territory impenetrable and inherently intractable. However, as we have hinted, it is possible to begin to answer these questions on many levels — physically, emotionally, mentally and spiritually — and to integrate our various answers and intuitions in to a more or less coherent philosophy. Since all things change, our ability to answer these questions, or at least our ability to come up with new answers, is also changing. One may even go so far as to say that our determined efforts to address the most difficult questions of existence are a sign that subconsciously we expect to find ultimate answers. The process of posing and answering questions presumably began at least as early as emergent language and self-consciousness allowed awareness of self-cosmos relationships. Since humans became the focal point of the stream of conscious evolution entering the river of unconscious evolution, there has been no looking back.

Countless philosophical, spiritual/religious and scientific commentators have offered their 'take' on reality. Some claim that their positions are based on logical deduction, induction, physical evidence, epiphany or divine revelation. In all such cases, individuals judge their insights (consciousness experiences) as worthy of communication to others, presumably because they have garnered some ostensibly meaningful insight into the nature of reality that they consider had previously eluded them and others. Thus, philosophers, spiritual gurus and scientists are among those most likely to convey messages that become part of our collective cultural heritage. For such messages to become part of the inherited perennial wisdom, they must resonate collectively and stand the test of time. However, unless we are extremely naive, or committed to rigid belief systems, we should know that science, philosophy and spiritual/religious beliefs are also subject to change.

Let us take some concrete examples. Ever since the Greeks gave us the seeds of what we loosely call modern western thought, the evolving mental-rational consciousness structure also known for two millennia simply as 'reason' (*logos*) has been much revered as the paradigm that gave us modern science, with all the geology, biology, psychology and other '-ologies' that *logos* spawned. (Some regard the meaning of *logos* as closer to 'wisdom' (*sophia*) — hence philosophy, anthroposophy, and so on.) As rationalism flowered, particularly in the eighteenth century we gave 'reason' a new and even more exalted label — 'The Enlightenment'. Even though most Enlightenment scientists were overtly religious men, their discoveries gave new insight into how the natural (physical) world worked, and religion began to take a back seat to science, at least in the realm of empirical study. Meanwhile philosophers like David Hume laid out exactly how they saw the new science weakening the influence of religion.[2] Scientists became the new 'priesthood', and by the twentieth century many were wearing white coats, working secretly in secure laboratories and speaking a complex new language involving a strange mix of traditional Latin and newly-minted technical jargon. Clearly a new or 'novel' paradigm was replacing an older model while retaining many of its conservative elements, including the status of the new priesthood, still ensconced in traditional institutions of higher learning with deeply religious roots. For example, among Cambridge colleges alone we have Christ's, Corpus Christi, Emmanuel, Jesus, St John's and Trinity Hall.

But modern science for all its technical accomplishments has not provided answers to life's fundamental, existential, metaphysical or spiritual questions. Hence at the dawn of the twenty-first century, scientists, especially those involved with medicine, are turning to alternative therapies which recognize humanity's spiritual dimensions and the danger of impersonal technology. In many circles the word 'enlightenment' has been co-opted to connote a trans-rational, intuitive consciousness associated with a new spiritual or pseudo-spiritual worldview that looks to reconnect with nature and the environment. Our aim is not to judge or measure rationalism against a new spirituality. Rather we wish to point out the dynamic flux in our linguistic and conceptual repertoire.

When the Harvard professor E.O. Wilson predicted a future battle for the hearts and minds of humanity between empiricists (scientists) and transcendentalists, he implied that, in his view, there would be only one winner — the empiricists, who will help us answer the ultimate questions by using scientific methods.[3] Wilson's 'battle' and 'one-winner' metaphors suggest that he does not believe in the advent of a new paradigm so much as a continuation of the old at the expense of the new. The twenty-first century model may be new but it still carries the label 'car' much like the twentieth century models. This dualistic prediction maintains the divide between what C.P Snow called *The Two Cultures*[4] without admitting the possibility of a 'third culture' — an integration of empiricism and transcendentalism under a novel banner. There is also no mention of why empirical science is so concerned about securing its future winnings. From a psychological viewpoint, this is a classic case of conservative, anti-novelty 'anxiety' about an impending shift that threatens to destabilize the status quo and bring about a paradigm shift, tipping the balance from the familiar, known to the uncertain unknown.[5]

In a dialogue between science and religion, the philosopher Michael Ruse adopts a modified but somewhat similar position. He admits that he does not 'think our science gets to ... the problem of consciousness' and this, he suggests maybe because 'a number of philosophers now suggest that perhaps the human thinking apparatus is just not strong enough to solve the problem of consciousness'. (Note here that the implicit assumption appears to be that the 'thinking' apparatus is the only one available.) He then, as if veering towards Wilson, states that: 'I don't mean that there is no solution of a natural kind ... I doubt that we can ever have proof that we cannot get to the solution, so we have to keep trying'.[6] This view could be interpreted as the self-evident statement that we cannot know whether there are unknown phenomena beyond our ken or, put another way, we cannot know whether or not God exists. Philosophers like Ruse are aware that science cannot prove the non-existence of God.[7] Evidently Ruse thinks Wilson is 'probably' right, that if a solution is to be found it will be an empirical, 'natural' science solution. We often hear scientists defend the position that science does not have all the answers, but at least it admits this limitation, deserving kudos for its inherent humility. But not wishing to be too humble, science usually reminds us that it is the best system available. This is what Owen Barfield refers to as the 'residue of unresolved positivism' — the notion that a rational explanation will always be found, and probably be the 'best'.[8] The entire argument could be repeated with equal force simply by substituting 'intuition', 'mind' or 'human consciousness' for science. Many of our most creative scientists admit that their breakthroughs often came when they gave up on a problem, went for a walk, experienced a revelatory dream or spiritual epiphany. All of this suggests that the very language we use to thrash out these problems is in a state of flux. Ironically, even if some hypothetical, ultra-rationalist empiricist used the most traditional scientific methods to make a huge 'consciousness' breakthrough that helped answer fundamental questions about our cosmic origins, there is no guarantee that history would not remember such a pioneer as a spiritual, consciousness genius or guru. This would be especially true and ironic if such a hypothetical pioneer were to describe such a breakthrough using popular science vocabulary like 'revelation', or insight into one of the great 'mysteries of the universe'. This brings us full circle to topics we shall discuss in the following sections: namely the influence of those who by their own admission or definition have gained 'cosmic' insight through epiphany, spiritual or non-rational, non-empirical means, when new consciousness insights simply emerged.

But first, another word about Ruse, who in his prolific writings frequently admits to being an empiricists and a Darwinian, albeit one with a commendably broad and open-minded grasp of evolutionary philosophy. No one should pontificate on evolution-creation debates without reading Ruse's thorough and well balanced treatments which show sensitivity for the historical currents that have influenced some to adopt positions that we today might find naive, bizarre or outright misguided. Ruse listens carefully to all sides without scorn. We also agree with his suggestion that 'we have to keep trying' although we might rephrase it, as intimated above, with the notion that 'we are instinctively compelled to keep trying'. What is not addressed so explicitly in Ruse's ruminations is that the psychological drive to answer the perennial questions, is more than a rational exercise of the 'thinking apparatus'. Ruse does not consider the work of Gebser, Barfield, Wilber, Thompson, Schad, Steiner, Verhulst and others who tackle the 'problem' of consciousness integrally with an appreciation for the 'ever-present' emergent properties of consciousness, which, incidentally are responsible for giving us new, if incomplete solutions, including those we then label as scientific hypotheses and theories. Indirectly, Ruse hints at the potential for the emergence of new consciousness structures to tackle new problems. This is the previously mentioned adage that we cannot solve old problems with the same thinking that created them. In a parallel to the 'thinking apparatus' metaphor, imagine a magic or mythic structure dialogue where someone were to say the 'human sleeping, dreaming and feeling apparatus is just not strong enough to solve the problem of consciousness'. Such discursive dialogue was probably unlikely, if not impossible in most ancient cultures. But leaving such speculation aside the point is that many insist that mythic (feeling), mental (thinking) and integral consciousness structures 'emerged' as part of the evolutionary process.

All this suggests that 'emergence' is an important part of the evolutionary dynamic. As noted in Chapter 7, commentators like Willis Harman speak of a global mind change already underway and they even suggest the teleological view that mind could be 'prior to brain' with evolution characterized by 'the organism's freedom to choose and by its inner sense of right direction' — a hypothesis 'congenial to the emerging view that consciousness has somehow to be factored into our total scientific picture of the universe'.[9] Such a view implies that humanity, and hence its evolution, is guided by some sort of moral compass — what our friend G.K. Chesterton referred to, with some reservations, as 'the doctrine that there is one great unconscious church of all humanity founded on the omnipresence of the human conscience' (see further in Chapter 9).[10] Other future predictions like those mentioned at the end of the previous chapter, suggest a type of critical mass trigger that would activate the Global Brain, when enough brains are interconnected. This neurological, computer-like metaphor is very similar to the notion of the noosphere proposed by Teilhard de Chardin.[11] Teilhard was a rarity among paleontologists, making bold predictions about *The Future of Man* and the evolution of consciousness.[12] In a hypothesis somewhat convergent with Russell's much later speculations about the global brain, Teilhard held that the 'psychic temperature' was rising as our network of communication (the noosphere) intensified in an absolute 'direction of progress towards the values of growing consciousness'[13] and 'the organic unity of humanity'[14] Teilhard boldly predicted Omega Man, 'a future being who would surpass *Homo sapiens* both intellectually and spiritually'.[15]

The evolutionary co-emergence of self and sensibility

In the previous chapter we looked at the different modes of thinking proposed by Wolfgang Schad. He considers that using the past to predict the present or the future may work for the physical (inorganic) world but may not work in the biological and psychological realm. In a similar vein H.S. Jennings notes that: 'As we look deeply into the past history of the world, it appears that there was a time when sensation, feelings, ideas and the like, did not occur ... A great class of things ... emerged in the course of evolution ... that could not have been predicted by computations based on what existed before.. [this] is what is meant by emergent evolution'.[16] Thus, states Jennings 'the emergent evolutionist holds that in the course of evolution there have emerged things that are new ... and ... not predictable from a knowledge of preexisting things ... New laws of motion, new methods of action, have appeared ... and peculiar mental states emerge'.[17]

Under the topic of emergent evolution, subsequently co-opted in another, strictly-physical guise as punctuated equilibrium (see Chapter 2), we can return again to that point in pre-history when self-consciousness first emerged among members of our species, and consider the psychological dimensions of emergence. This was a momentous 'phase shift' in consciousness structure, which recapitulates in our individual development at a very early stage, and we can speculate that it is intimately tied to the emergence of language, awe and spiritual-existential ponderings in our species history just as it is in individual development. Many commentators consider that it is self-consciousness and language that make us human, and leave it at that. These commentators are almost certainly correct in this assertion. After all self-consciousness and language allow us to recognize these faculties in the first place, and there is no reason not to cherish and value them highly. What we find most intriguing, however, is the momentous impact on evolution, and how this shift repositioned our role (or more strictly how we began to see our role) in the world.

We touched on this before (Chapters 4–5), concluding that before we humans became self-conscious the world literally did not exist for us psychologically. This raises the fascinating question of what we perceived before we 'came to our senses' — before we became identified with our physical bodies. Barry Long has proposed some very interesting answers, which may at first seem provocative to the rational mind. But increasingly we have found scholarly support for the gist of many of Long's ideas. Long was born in Australia in 1926 where he became a journalist. Subsequently, in his thirties , as a result of his spiritual experiences, he went to India as a spiritual seeker. He later became a recognized 'spiritual master' or 'guru' and authored sixteen books translated into a dozen languages. In exploring his insights we find they are remarkably consistent with all we have gleaned from the broader field of consciousness studies.

Simple logic tells us that if our ancestors 'lived in' the unconscious realm they obviously functioned and 'survived' perfectly well. Moreover, they had no worries about survival, as concepts like death, did not exist for them. They were capable of functioning quite unconsciously and effectively, just the way many of our organs and physiological systems continue to operate today. Long states: 'Primitive man lived mostly in his vital double ..[he] was not an individual as we today can experience ourselves. Primitive man experienced the world as himself — the consciousness centre of the vital or psychic brain. When primitive man needed food he knew psychically in which direction to go and hunt for it. ... without exterior sense aids ...'[18] Here we might ask whether there was any need for discursive language?

This unselfconscious, internal worldview is evidently consistent with the zero dimensional archaic structure suggested by Gebser (Chapter 5), as well as the perspective envisaged by Edwin Abbot for the zero-dimensional realm of 'pointland' where it is impossible to conceive of anything 'other' than self — where it is impossible to perceive dimensions of 'outer space'. It may seem paradoxical to 'experience the world as self' but not be self-conscious, but as we established at the outset (Chapter 1) one can only consciously perceive the 'self' in relation to something 'other'. Without the other there is no self-conception and vice versa, and, we might suppose, little or no need for language.

If we try and place the emergence of self-consciousness in the schemes of previously mentioned scholars, we might conclude that the shift arose as Gebser's archaic consciousness structure transformed into the magical state. This does not mean that every shift between the five Gebserian structures involved phase shifts equal in fundamental importance to the emergence of self-consciousness, but we argue that there is growing, and internally consistent evidence for at least two major shifts. The first broadly equated with the Id-Ego transmutation and the second with the shift from Ego-Superego. For example, according to Barfield, transformations take place between three states: original participation, separation and final participation. Although the schemes have different labels, the cycle of unity, separation and reunification is common to both these. This enables us to propose the following scheme (Table 8.1).

In our quest for internal consistency and agreement with organic principles we note that this organization also resonates with the inherent twofold 'structure' of existence: namely life and death, which can equally well be characterized as a threefold structure: before existence, life and after life (past, present and future) or even as we light-heartedly suggested in our introduction to this book:

nowhere — now here — nowhere

We use this schema to argue that human history provides overwhelming evidence that just as we 'fell' into the world by undergoing a psychic separation from the cosmos, so it is possible to 'reunite' or reintegrate with our cosmic roots. This psychic separation and reunification is what gives us our spiritual or religious nature and all its attendant psychological dilemmas. As Rudolf Steiner often said it is part of the human condition to face problems. The miracle of birth creates a new individual soul, but it is accompanied by the anxiety of separation, and the challenges of relationship with the world.

By analogy, humanity's collective fall into self-consciousness creates a separation from the cosmos and its unconscious wisdom, leaving us to figure out our relationship with the mysterious domain of the unknown that we intuit but, by definition, never consciously knew in the first place. But our very instinctive awareness of the unknown drives us to ask the big metaphysical questions and explore, and as far as possible make known, not only the newly-opened world of experience, but also the ever-present, but 'intuited' unknown. For this reason, we submit, that cosmic consciousness epiphanies and other varieties of spiritual or mystical experience are normal manifestations of our ever-present relationship with our psycho-spiritual origins. We shall deal with these further shortly, because unlike the phase shift into self-consciousness which is obscured by prehistoric mists, the mystical experience has been extensively documented.

Table 8.1. *Two major phase shifts in consciousness (self-awareness and mystic experience) suggest that fundamental psychological dynamics underlie our religious and spiritual sensibilities and are an integral part of our conscious evolution.*

Id	Ego	Superego
Original participation	Separation	Final participation

Archaic	Magical	Mythical	Mental-rational	Integral

	Separation and the origins of self-consciousness and religion.		Mystical reunification.	

Let us return first to what it really means to 'fall' into self-consciousness. Barry Long points out that, before becoming self-conscious one could not even be aware of pain or death:

Primitive man felt pain. But he did not know whose pain
it was. He had a kind of broad tribal identity, but not a
communal existence in which he shared his female ... she was
neither his nor anyone's. Neither she nor any other thing he
needed ever appeared in his feeling world or consciousness
until he had the feeling of desiring her or it. Between those
feelings female had no existence for him as male had no
existence for her. ... Rival tribes fought for what they needed-
without any sense of victory or loss. Primitive man's battles
were actually fought within himself.. he merely registered the
feelings ... as instantaneous in its communication to him as
our sense perceptions are today. The difference was there were
no images, no concepts, no forms. He had not realized his
sense, his physical body, so his body could not exist for him
[as] for a hypothetical observer.[19]

This suggestion is similar to that made by Vitaliano when he states
that: 'dualism is the act of severance, cutting *(con-scire)* the world
into seer and seen, knower and known ... with the occurrence of the
primary dualism, man's awareness shifts from the non-dual universal
consciousness to his physical body'.[20] Steiner suggested something very
similar (see Chapter 4) when he implied that before humans developed
memory, they could not attach importance to their deeds or have any
Ego sense of 'victory or loss'.

Paradoxically, although today we have rich psychic lives full of
thoughts we somehow find it difficult to envisage a world where
the psychic life was dominant and divorced from a physical sense of
self, although our dreams represent a very different psychic domain.
Consider that no less an authority than Carl Jung has encouraged us
to remember that: 'It is almost an absurd prejudice to suppose that
existence can be only physical ... We might well say, on the contrary,
that physical existence is a mere inference, since we know of matter
[only] in so far as we perceive psychic images mediated by the senses'.[21]

As consciousness shifted, Long maintains:

... man was no longer content to remain a part of the
unconscious. ... darkness was dissonance ... vague stirrings
urged him upward. Outside in the light, on the earth, the
truth awaiting him was his physical body. For a long time
... physically realized men were a rare phenomenon ... able
to retreat into the psyche and communicate to the others
coming up behind them. To the mass of self-unconscious man
they were literally beings from outer space. ... A physically
realized man was a very privileged specimen ... [they] were the
mythical and legendary gods. He could communicate with
his developing companions in either the outer ... or the inner
world ... put suggestions into him, frighten him ... induce him
to worship and serve ...[22]

This perspective on the emergence of self-consciousness has
interesting implications. Long suggests that:

no rational line could be drawn between fantasy and physical
reality. Men lived in both worlds with the gods — the
realized men having the best of both. The gods were men.
They hated and loved and warred among themselves ... hazy
psyches were used by the gods to wage fantastic wars against
each other ... It was like a group of particularly privileged
people waging wars of dreams.[23]

Before the ancients emerged from their unconscious archaic world
they had been psychologically immortal, the concept of death did
not exist. But, as they became aware of their bodies, attachment
naturally increased, and with it emotional pressure. Physical 'death ...
the fear of losing his body became obsessional. [Man] had to find ...
an escape ... another world that did not end in death.. Tomorrow was
his answer. The world's first act of ignorance ... [because] ... the truth
is immortality'. Nothing had changed in the cosmos, only mankind's
view of it.

Today it is almost inconceivable to us that once it [tomorrow]
did not exist ... the forming of human nature ... was no more
than the complicated outcome of a very simple and purely
psychological device for escaping from emotional pressure. By
creating ... problems around a supposed future ... he was able
to keep his mind ... from having to face up to the big problem
of today — death. So death was projected hopefully as an
event of the future. [So] the ... problem ... was eliminated ...[24]

These interpretations pivot on a fundamental question which
seems to be at the core of most of our religious and spiritual
traditions: namely, are humans spiritually and psychologically
immortal, even though their bodies perish? Long says yes, and as we
have discussed, life, of which we are a part, is effectively immortal.
From the time that the first archaic humans and Neanderthals
started burying their dead, it is clear that humans have entertained a
belief in an afterlife and the immortality of the soul. Even the most
materialist anthropological interpretations agree that burials and
countless other funerary rituals speak to these beliefs, even if those
acknowledging such beliefs do not share them. As Ruse points out,
it was only in the sixteenth century that western thinking began
to dispense with the notion of the world as an organic rather than
a mechanical entity.[25] Ironically, as quantum physics informs us,
the mechanical paradigm is already as suspect and 'moribund ' as
its characterization of the world. Meanwhile the Gaia hypothesis
revives the notion of an organic world (biosphere) which, being as
long-lived as life on Earth, is therefore essentially immortal.

Given these ruminations it seems logical to ask whether widespread
spiritual and religious beliefs in immortality, are not based on our
instinctive and intuitive awareness of the immortality of the life
force in which we all partake. Is Long's reference to the unconscious
'psychic web of the biosphere' different in principle from conventional
paleontological wisdom which views the history of life on Earth as a
complex, 3.5 billion year old co-evolution of ecosystems and behaviours
involving hundreds of millions of species?

Philosophically we need to ask why we would not, as Jung suggested, take our psychological experiences and those of our ancestors as concrete manifestations of reality, rather than patronizingly branding them as imaginative delusions. Barry Long takes just this literal approach when stating that:

> The Greek myths and Hindu traditions tell of gods and demi-gods walking the earth ... yet no one has ever seriously suggested that this was precisely how it was ... Paradoxically, our civilization has retained these averred scriptural and mythological fairy tales as an essential part of its reservoir of learning and thinking. Even ... C.G. Jung saw the energies of the myths acting outwardly through the subconscious but still failed to get back behind the symbols to identify the actual event in time that originated them ... Neither the Greek myths, the bible or the early Sanskrit texts ... mentions that the accounts do not stand for exactly what they say. .. Central ... to all religious traditions are the unqualified references to the gods and demi-gods with miraculous powers participating in human affairs ...[26]

Fantastic as this may at first sound to the modern rational mind, we would do well to ask why so many of our modern movies develop precisely these themes. What is a movie if not an external projection of the human psyche? Why, down through history, have we believed in possession by demons and spirits? Why in modern psychological language do we still use the metaphor of demons for unconscious issues that upset our psychic equilibrium, and perturb our physical, emotional and mental bodies? Is it simply because the spirit world is also real?

Perhaps it is easy for the rational mind to dismiss demons and movie monsters as playful creativity, but it is less easy to explain why serious anthropologists study the shamanism, animism and telepathic abilities of the Pygmies, !Kung, Hmong and countless other cultures. Would we take their experiences and deeply held beliefs in spirits seriously if they were ultimately judged by researchers as delusional worldviews? As Barfield pointed out (Chapter 5) there is not a shred of evidence that animistic, magic or mythologically-oriented cultures make up explanations for natural phenomena simply because they cannot interpret them rationally in the same way we do. In general, such cultures are describing their actual experiences. Long suggests, closely echoing Barfield:

> ... man, before becoming a self-conscious being ... was linked inwardly to all things as a psychic or feeling ethos. He was like our brain which is the centre of our afferent and efferent nervous responses. What thought is to us feeling was to him. He was the optimum of clairvoyant and extra-sensory communication except that he had no sense, no individuality, no personalized life.[27]

Shamanism provides considerable support for this claim of clairvoyance and extra sensory communication. As many anthropological treatises suggest, shamanism can been described as the world's oldest religion.[28] This label has also been applied to mysticism, but while both deal what we might loosely call altered non-mental states of consciousness, they are not the same. Shamanism is highly animistic and totemic. The shaman is capable of going into a trance, an altered state of consciousness in which he/she is able to commune psychically with the dead, animals and other spirits. In this trance or 'ecstatic' state — literally meaning to 'stand out side' (*ex stasis*) of normal consciousness — the shaman is literally a bridge between two worlds, that is, the supersensory trance world of the unconscious (animistic nature) and the sensory world of conscious awareness. The shaman is a psychic traveler who is still capable of bridging the consciousness divide between the archaic collective, which Long has called 'the psychic web of the biosphere, ' and the world of individualized, or individualizing self-aware humans.

One might think that if humans evolved in this world, progressing from 'bottom up', to ascend the 'great chain of being' from fish and amphibians, to reptiles mammals and primates, that they would have no experience other than a physical grounding on planet Earth. But paradoxically any lingering traits of pre-human consciousness, would probably be quite different and of 'another world' of experience.

Psychically, however, from the earliest days of shamanism, and ancient written cosmologies the message is of origins or descent from the 'top down'. (See Table 7. 1.) It seems humans are acutely aware of 'other' worlds. One interpretation of the paintings and stencils of hands from Paleolithic caves dating back some thirty thousand years is that they indicate 'a reaching for what lies beyond the cave, beyond our world in the realm of the dead and the ever after'. This interpretation of people reaching for the world beyond was made, not by modern anthropologists, but by a San tribesmen (Bushmen) and an Australian aborigine who were invited to view the famous hand paintings at Gargas cave in France.[29] One could no doubt write a book about 'waving' and 'hand signals' as symbols of communication and relationship with the 'other'. It is appropriate that the 9,000-year old hand paintings from Cueva de las Manos, Patagonia, which wave at us from our book cover, and Figure 8.1, represent a UNESCO World Heritage site. Such symbols of humanity's reach for other worlds — other dimensions of consciousness — are indeed quintessential facets of our psychological heritage. Such hands reach across time from *Homo habilis* ('handy man') to *Homo erectus*, the creative maker of the first aesthetically-pleasing 'hand axes', and to artistic *Homo sapiens* whose worldwide creative handiwork brought to life a tradition of colourful galleries of hand paintings waving to us from deep time.

Could it be that the relationship between hand paintings and caves is also symbolic of reaching across the liminal threshold between seen and unseen worlds? Certainly the cave is a very intriguing and familiar motif. We need only think of Plato's cave where we see only shadows cast from another world, and Jung's dream of descending through the basement of his house to ever-deeper levels with ever-older cultural artifacts. Most contemporary religions seek psychological or next life access to the world beyond, to the ever after. And many religions hold that this other world is the one from whence we came and where we ultimately belong. Even a neutral or non-religious interpretation allows that we cannot know or experience all there is, and that we must therefore acknowledge psychological or psychic dimensions beyond the reach of present consciousness.

Figure 8.1. Nine thousand-year-old hand paintings from Cueva de las Manos, Patagonia, may be interpreted as humanity reaching for another world. See cover for colour rendition.

When we speak of 'coming into being' our language betrays the idea that we came from 'somewhere else'. Is this sense of another world the inevitable by-product of becoming self-conscious, which requires that we recognize the other? Did we descend from on high, from the realm of the Gods, or arise from amoebas and apes? We have already asked what it means that we use both evolutionary metaphors interchangeably. Descent and ascent are both metaphors for an innate sense of evolutionary movement or direction. The question of origins is inherently paradoxical. We only became conscious of origins when we became self-conscious: so how can we ask or know what existed before we originated (existed) in our present world of experience — before we came into being? Why do we contemplate the possibility of an existence 'beyond' this experiential world? Why postulate a hypothetical realm outside our space-time experience? Is it mere imagination, or is it because we once had such unconscious 'experience' and indeed continue to function on this level in sleep and deep physiology?

Marcelo Gleiser points out that there are two polar cosmogonic myths: one speaks of an eternal universe with no beginning, the other postulates an abrupt beginning.[30] The fact that we can ask what existed before the beginning means that we can contemplate something other than a defined beginning. Practically all cosmogonic and cosmological myths and stories 'begin with' or involve some form of creation or origination to account for our existence and our ability to think about existence. Our mental preconceptions tend to make contemplation of deep time (and space) an intractable puzzle. How can we contemplate a universe that has no beginning and no spatial boundary? On the other hand it is equally impossible to contemplate a universe with a boundary and beginning without asking what is beyond and what was before. One partial solution to the dilemma is found in worldviews that allow for atemporal and well as temporally constrained consciousness structures. There was a period in our infancy, and when our species was young, when such time-haunted questions were literally unthinkable or of no consequence. This timeless consciousness structure is not lost, and as we shall see, it is still accessible.

In the beginning was the word, and the word was sacred

In considering the implications of becoming self-conscious and aware of our bodies, it appears we inevitably 'fell' into a new and magical sensory world of nature and cosmos. To use a previous metaphor, it is almost as if we tuned into a new channel, and literally discovered a new world. As a result a new and innocent sense of awe and wonder emerged. We might call this religious or spiritual sense a natural by-product or corollary of self-consciousness. Such a sense could only be absent if we failed to find nature interesting and intriguing to our sensory perception. The origin of language is so elusive and difficult to explain, rationally, that in the 1860s and 1870s several learned linguistic societies banned any further speculative contributions on the subject. However, more than a century has passed since then and a few insights are worth mentioning.

In the 1870s, as modern anthropology developed, scholars first began, rather patronizingly to consider that our so-called 'primitive' ancestors had consciousness structures different from our own. Pioneers like Max Müller, a specialist in language, considered the origins of religion and language to be deeply intertwined and he developed what he called a 'science of religion' and attempted to divine how the ancients interpreted the world, by interpreting ancient texts.[31] Müller accepted that human form (anatomy) evolved from animal form, as Darwin had suggested in his *Origin of Species*. However, Müller refused to accept that human consciousness and language had evolved biologically from animal consciousness. Müller even explained this to Charles Darwin directly in 1873, to which Darwin is reputed to have replied: 'You are a dangerous man.'[32]

In *The Biology of God*, the late Oxford biologist Alister Hardy suggests that differences between animals and humans relate, in part, to the differences between what the science philosopher Michael Polanyi, calls *tacit* and *explicit* knowledge.[33] Tacit knowledge is gained subconsciously from experiencing the world around us. Explicit knowledge is formulated in words, maps and symbols as an explanation:

it comes, in a sense, 'after' tacit knowledge. Thus the difference is similar to what Barfield calls figuration (tacit knowledge) and alpha thinking where we collectively and 'explicitly' compare our representations using language. Thus, animals also have tacit knowledge but not an explicit means of linguistic expression. However, as a biologist, Hardy believes that animals and humans share many emotional traits and behaviours. Darwin, of course, elucidated these profound similarities in his book *The expression of Emotions in Man and Animals*. Likewise, Hardy also sees the aggression-submission behaviours of many vertebrate animals, especially mammals, also expressed in humans. Thus, he believes that our sense of a higher power, to which we submit, or pay credence, is fundamentally a biological emotion. What might this have meant for ancestral humans first coming to their conscious senses, and beginning to try and make explicit sense of a world previously known only subconsciously or tacitly? According to Hardy it meant a willingness to submit psychologically to powers greater than ourselves.

In contrast to Hardy's ideas, which put biology before psychology, Müller's approach has been described as 'intellectualist' in the sense that it tries to understand how the ancients invoked 'animist' theories, which inferred that primitive peoples invested all nature and life forms with soul, and believed that spirits of departed ancestors led their own independent existence.[34] In the nineteenth century, this was widely regarded as primitive superstition that predated what Krishna calls the 'lofty' conceptions of Gods and deities that developed later.[35] But this élite view is largely a flawed Victorian projection by smug Europeans. Müller nevertheless did address the evidence that the ancients did actually develop lofty conceptions of large intangibles like sky, sun, dawn and the heavens and were by no means initially fixated on animating tangible objects such as rocks, plants and animals. Müller also roundly demolished the patronizing theory of 'fetishism' as a archaic religious impulse among so-called primitive tribes. He pointed out that it was eighteenth century Portuguese sailors who projected their own fetishism on to the African natives of the Gold Coast. It was the sailors 'themselves [who] were the worst of fetish worshippers' with all manner of Catholic icons.[36]

Thus, Müller's work contains many valuable lessons. As a linguist he realized that to understand ancient cultures, one must decipher their meaning, and not project our language and meaning into their world. Barfield became a leading exponent of this philosophy in the twentieth century when, as his book *Speaker's Meaning* clearly indicates, he stressed that the study of language, meaning (and hence religion) are inextricably linked to the evolution of consciousness.[37] To infer a childish or primitive way of thinking is merely to project our abstract notions of primitiveness into a quite different psychological world.

With the foregoing introduction, let us turn to a fascinating study by Ernst Cassirer to find a deep resonance with many of the sentiments stressed above:

> ... the whole realm of mythical concepts is too great a phenomenon to be accounted for as a 'mistake' due to the absence of logically recorded facts ... Reason is not man's primitive endowment but his achievement ... Philosophy of mind involves much more than a theory of knowledge; it involves a theory of pre-logical conception and expression.[38]

Cassirer explores meaning by stating that 'the notion that ... the potency of the real thing is contained in the name ... is one of the fundamental assumptions of the mythmaking consciousness ...'[39] He goes on to say that:

> language could not begin with any phase of 'noun concepts' or 'verb concepts' but is the very agency that produces the distinction between these forms, that introduces the great spiritual 'crisis' in which the permanent is opposed to the transient, and Being is made the contrary of Becoming.[40]

This is surely precisely what is meant by falling from the eternal, atemporal and permanent world of the unconscious into the transient, temporal world of time. Cassirer agrees with the German scholar Hermann Usener (1834–1905) that:

there have been long periods in mental evolution when the human mind was slowly labouring toward thought and conception and was following quite different laws of ideation and speech. Our epistemology will not have any real foundation until philology and mythology have revealed the process of involuntary and unconscious conception. The chasm between specific perception and general concepts ... is so great that I cannot imagine how it could have been bridged had not language itself, without man's conscious awareness, prepared and induced the process.[41]

So Cassirer poses a very pertinent question highlighting the importance of historical linguistics (also known as 'philology'): 'Is there any other line than the history of language and religion that could lead us closer to the origin of primary linguistic and religious concepts?'

Blending his ideas with those of Usener, Cassirer asks how the complex rhapsody of perception, also experienced by animals, is converted into a coherent system of meaningful linguistic laws. He says we must compare primary linguistic forms with mythic ideation, not with logical conception. Mythical thinking is 'captivated and enthralled', resting only in the 'immediate experience' without comparisons with the past and future. So 'the form of language should not be traced back to any form of reflective contemplation'. Cassirer cites *Hamann's dictum* that poetry is 'the mother tongue of humanity'. Speech is rooted in the poetic aspect of life, not the prosaic; in the primitive power of subjective feeling, not the 'objective view of things'. Here again we find great internal consistency with the work of Barfield who equated poetry with inspiration (perception) and prose with conception (see Table 5.2).

So how is the sound of speech transformed from an emotional utterance to one that has conceptual meaning? Once the 'word' is uttered, 'a turning point has occurred in human mentality', inner subjective excitement becomes speech, representation, myth. The spoken word calls something from the total sphere of consciousness, sets a seal on things and gives them form. In rudimentary form, the animal's alarm call or mating cry, is an emotional cry. Likewise the child's early utterances are an expression of subjective feeling in response to the experience of the external world. Often they are what we may, rationally and 'after the fact', define as spontaneous utterances of pure joy or disgust.

As an instinctive creation, felt deeply as a vibration, the word has magic properties. In all cultures we find that veneration of the word is a big deal. Ptah — the Egyptian creation god — is a spiritual being who can think and speak the world into existence using the heart and the tongue. This creates Cosmos from Chaos. Language is a creative act as fine literature attests. The Vedas say 'the word is the Imperishable, the firstborn of the eternal Law, the mother of the Vedas, the navel of the divine world'.[42]

What's in a name?

The name is a proxy for both gods and mortals. Without a legal name a Roman slave did not exist, and to this day tyrannical regimes suppress minority languages in an attempt to eradicate other cultural identities. Algonquins said a person with the same name was another myself. And many cultures hold that you can frighten away death with a scary or false name. Even today it is, on the one hand, taboo to speak politically incorrect names, or take the name of God in vain, while on the other using another's name is key to fraud and identity theft. Language is essential to identity and makes community existence possible.

Cassirer says that the origin of language and religious sensibility go back to the sense of undifferentiated, supernatural power (*manu* in Melanesia, *manitu* in Algonquin or *wakanda* among the Sioux) which early missionaries and anthropologists referred to, oversimplistically as the Great Spirit (or God). The real meaning is not a thing (a noun, a being or a power) so much as a mystery, a term with sacred, special, magic, supernatural, delightful, meaning distinct from the profane, or mundane. The terms could be used as noun, adjective, verb or adverb. *Manu* and such terms are 'so protean' as to 'not be susceptible of translation into the more highly differentiated language of civilization'.[44] Nevertheless, as we shall soon see, this mysterious and sacred sense seems to have been a perennial source of power and inspiration in most, perhaps all cultures, since the dawn of human consciousness: it is what renowned anthropologist Emile Durkheim called the means of communication which makes 'a man who is stronger' allowing him to feel 'within himself more force, either to endure the trials of existence or to conquer them'.[45]

I-THOU and ME-WE

The self-conscious dualism of me-thee, us-them and I-thou is deeply ingrained in our cultural tradition. It is also a fundamentally religious as well as a social concept. Thus, it is important to consider the needs of the other, as in the Golden Rule. In many spiritual traditions the relationship of the individual to the divine is often expressed as an I-thou relationship.

More prosaically, but none the less poetically, Norman Mailer recalls an occasion when students asked Mohammed Ali for one of his famous poems. His spontaneous response was 'Me-We.' Mailer suggested that Ali had created the world's shortest poem, replacing the flea poem: 'Adam had 'em'.[43]

Giving a God many names (polynomy) is a sign of enthusiastic feeling (Allah has ninety-nine names, some Egyptian Gods had ten thousand.) Speech both divides (many names) and generalizes (unites). So early in the evolution of language, speech helped to divide the unity of the cosmos into the 10,000 things referred to in Buddhism. But language can also transcend itself and say that ultimate reality is beyond linguistic explanation. It just is what it is: the Hebrew God said 'I am that I am'.[46] Mystics like Meister Eckhart have declared that whatever you say God is, he is not that. Likewise, Taoism states: Whatever you say the Tao is, it is not that. God is hidden — 'a secret to his creation'. Absolute reality cannot be expressed through any analogy with things or names of things. Hinduism says that the Self (Atman) can merge with the Godhead (Brahma — the inaccessible: the 'it is not so'.) Thus we have two 'ineffables' of a different order, one of which represents the lower limit of verbal expression, the other the upper limit.[47] In our judgment, this distinction is very much the same as that made in Table 8.1 in distinguishing between the phase shift first into, and later beyond, self-consciousness.

Moving beyond the lower limit to language and naming of things, helps us create identity and a sense of self, and move out of Chaos into Cosmos. Humans take physical possession of the world. This is the spiritual achievement of monotheism, and it seems no coincidence that the shift from polytheism to monotheism is intimately associated with the development of self-consciousness and the strengthening of Ego. Once we see ourselves as independent beings we are also capable of giving God a well defined and coherent identity. These seem to have been a prevalent trend in Keck's patriarchal epoch II when the term 'father' came into widespread use. We may even follow these developments sequentially. Biblical narratives indicate that the western God image underwent a steady humanization in the course of history as the dialogue between early Jews and their often wrathful, jealous God incites conflict and forces the development of conscience (consciousness). Jung argued that when man, represented by Job, recognized God's inner antinomy, this knowledge of God's inner contradictory nature allowed Job (man) to attain a divine numinosity. Thus, man steals a piece of moral high ground and the God image descends into humanity, incarnating partially, in a good and heroic Christ, who like other demi-gods was part God and part human. When taken another step, as discussed below, we encounter the mystical realization that 'God and I are one' and the descent of God into man becomes complete. This notion is succinctly summarized in the ancient aphorism attributed to Athanasius of Alexandria (AD 293–373): 'God became human so that man might become God'.

Returning to the deeper meaning of language, Cassirer ends his thesis with a discussion of the power of metaphor which he regards as the intellectual link between language and myth. What metaphor does is identify one object with another: for example 'the sea is a desert', 'life is a journey'. Some say speech creates myth (through the use of metaphor). Others say the opposite: that myth created speech. This chicken and egg debate is largely irrelevant. 'Language and myth stand in an original and indissoluble correlation to one another, from which they both emerge ... gradually as independent elements' as the result of the 'heightening of simple sensory experience'. How do we get back to the 'radical metaphor' that came before things were named, given meaning and available for comparison? One can infer an impulse to raise an impression from the realm of the ordinary (profane) to the level of the holy (sacred). Usener says: 'people do not invent some arbitrary sound complex to [denote] a certain object ... The spiritual excitement caused by some object ... the self ... encounter with the not self ... naturally strive[s] for vocal expression'.[48] Our language speaks to the 'spontaneous outburst of joy' — to the 'squeal of delight'.

The story of Helen Keller, the child who was blind, deaf and mute before she acquired language, provides what we might call irrefutable evidence for this 'spiritual excitement' mode of language acquisition. Though her handicap could have prevented her from ever developing human language, through painstaking work her governess helped her to do so. She was able to learn the concept of water as her governess held her hand under the water and spelled out the letters 'w-a-t-e-r' with her fingers. 'I knew then what water meant. That living world awakened in my soul. Everything had a name and each name gave birth to a new thought. I *saw* everything with a new light'.[49] Note here that *seeing* is equated with the super sensory ability to awaken to or 'imagine' the concept. It has nothing do to with sensory visual seeing. Thus, without language there is no meaning, no world in the way we humans 'mean' it. It is for this reason that feral children never become human in a linguistic or a cultural sense.[50]

Mythic perception only slowly extracted special significance or meaning from the undifferentiated mass of experience, the rhapsody of perception. So in ancient times one thing could substitute for another. One could have power over a person by using their name, a nail, a piece of hair, spittle, a footprint or something that person had touched. We referred to this as 'sympathetic magic' in Chapter 5. This is the equivalence principle that is the essence of metaphor. Two things quite different to the mental mind: the collective 'herd instinct' and the animal God Pan had the same name and same dim meaning. Only after many things had been named could they be reorganized in differentiated, groups with individual meanings. So herd instinct (panic) became one thing and the mythical God Pan another. Likewise it is well known that Greek words like spirit (*spiritus*) and Latin words like *anima*, had broad meanings pertaining to physical breath, and by implication 'life force', vital principle or soul. English and other European languages have evolved and extracted many meanings from these roots: respiration, inspiration, aspiration, spirituality, in the former case, and animal, animation and 'anima', meaning soul, in the latter. Language first evolved to represent direct 'qualitative' sensory or emotional experience. As it did so it became an abstract substitute for the experience, an intellectual process of distancing the word from the emotional experience. Finally however, through poetry and artistic expression language can retain or recapture its original creative power. This is a type of freedom from 'hard' mythic bonds. A liberation of the mind as Owen Barfield says: as thinking separates from feeling.

Sin, sacrifice and the loss of innocence

Before we expose our psychological sensitivities to the somewhat disturbing world of animal and human sacrifice, let us remember a few biological facts. As noted by Alister Hardy, humans have a history of hunting. This makes them more carnivorous than almost any other primate. He believed that hunting behaviour was a major social factor in developing group communication, and being able to dominate and kill large animals. As discussed in Chapter 5, as humans became aware of their mortality in the early (archaic and magic) phases of development, so awareness of the act of killing an animal also became a part of conscious and psychological experience. Humans can do harm to, or be harmed by, the same natural order of which they are part.

Since the original separation of the self-conscious individual from the unconscious psychic web of the biosphere is lost in the mists of prehistory, contemporary infant experience provides the best alternative psychological analogy. When the child separates psychologically from the mother, in a very real sense it sacrifices part of itself — a part of what was once itself. This is what psychologists may call the fearful 'schizoid-paranoid' stage of development, the inevitable result of a psychic split that accompanies separation. This loss of innocence is accompanied by a sense of resentment directed from the newly emergent self towards the 'other' (parent). Of course in infancy it is an unconscious 'feeling' — not a thought, but still real and capable of generating conflict or guilt because the child still depends on the other (mother) while also resenting her transformation into a separate object. On another level, the infant self also resents the dependency on the other as it begins to separate and become independent. The psychic or schizoid split is organically re-integrated if the mother and infant have a healthy (non-abusive) transitional relationship where the mother acknowledges her responsibility to simultaneously nurture the child's growing independence and continued dependency. The child next reaches the 'depressive' stage when it comes to accept its own ambivalent feelings about dependence and independence.

Grotstein outlines these stages and their implications showing how loss of innocence (and resentment on the part of the child) leads to other dynamics such as a resentment towards the father (Oedipus complex) and the sense of original sin.[51] Original sin which can be equated with animal nature is equivalent to the feeling of guilt, the reciprocal of innocence. The child 'should not' feel resentment towards the mother or father on whom it depends for nurture, while also feeling the need for independence, which it cannot adequately express. As this ambivalence is impossible for the child to comprehend cognitively, it is up to the parent to help strike a balance so that feelings of guilt or naive innocence are not repressed or exaggerated (requiring psychoanalytical intervention later in life).

The whole gory tradition of human sacrifice can be linked to the psychological dynamics associated with loss of innocence and the feelings of guilt (animal nature conflicting with emerging human nature) and impurity that surround it. If the victim of abuse (= an unhappy balance between parent and child) feels undue guilt for not having earned/reciprocated the parent's love and support on which it depends, it may become a martyr, offering itself in sacrifice as atonement to the other. Conversely, the offspring may act out hostilities towards the parent (schizoid-paranoid syndrome) in an attempt not to internalize the guilt. In either case, it creates psychic conflict. If the parent abuses the child, it sacrifices the child as punishment for its innocence. Nevertheless the sacrificial victim is always (by semantic definition) sacred because he/she serves the purification purpose of atoning for the sin/guilt of the other.

The prevalence of sacrifice in ancient cultures, considered horrific to most modern sensibilities is, according to Grotstein, an indication of a psychological immaturity among the ancients, stuck in — or passing through — the infantile and unconscious schizoid-paranoid stages of development. The profound psychological insight attributed to Jesus when he said, 'Father, forgive them for they know not what they do' has been interpreted as the passage in which 'we are given the first definition of the human unconscious in human history'.[52]

So it seems human evolution has involved a journey through these various psychological stages, recapitulating the psychic ontogeny experienced unconsciously, subconsciously and consciously by each individual. Along the way psychological experience has been transformed into powerful religious symbolism, which in has turn had the same powerful influence on society, cultural morality and law that the individual dynamics have on the individual. This means, in our opinion, that we cannot separate the origin of self-consciousness and language from the origin of religion, and our ultimate psychological and spiritual origins. Erich Fromm's noted book *Fear of Freedom*, speaks to the psychological emancipation that comes from breaking with the more authoritarian traditions of institutionalized religion and thinking that is so often generated and sustained by guilt.[53] Fromm speaks for the freedom, and concomitant responsibility for self-realization which comes from a more mature psychological appreciation of the complexities and potential of consciousness.

Long before certain cultures realized the psychological freedom of which Fromm speaks, many had dispensed with human sacrifice, substituting animal sacrifice as an alternative. In the biblical tradition, this major psychological shift is told in the iconic story of Abraham and his son Isaac.[54] As Abraham is on the point of sacrificing Isaac to God, God speaks to Abraham, rewarding his obedience by providing a goat (scapegoat) so that his son can be spared, and all relieved of potential guilt. There is no easy escape from the psychological effects of living in a world where the cosmos (gods) inflict pain and death. For the ancients, the experience of birth and death as cosmic phenomena may have stirred awe in the supernatural forces (*manu, manitu, wanaka*), creating an urge to revere and emulate phenomena as sacred manifestations in the profane world — allowing what, in the case of sacrifice, Dudley Young calls 'an offering back to the gods of their mutilating powers' ... encouraging us to believe that sacrifice is somehow 'natural' and 'divinely ordained'.[55] The Romans went in for animal and human sacrifice on a huge scale, and the Coliseum, awash with blood, became the enclosure where humans had an uneasy and tenuous power over the wild, savage and unconstrained chaos of nature outside. Bizarre as it seems to apply the modern phrase 'doing God's work', in a strange way the Romans believed they were appeasing the gods.

As Young notes, every culture has its history of hunting and most also have well documented histories of sacrifice. Few cultures go to the extremes of Jainism where the ground is swept clear of insects lest they be inadvertently killed. Whether humans are killed on sacrificial Mayan altars, in Roman amphitheatres, as retribution for crimes against society, or in supposedly just or unjust wars, and whether animals are killed in the hunt or again, as retribution for crimes and threats against society, humans emulate the supernatural powers that periodically cause blood to be spilled. No deed goes unpunished, and the psychological fallout manifests in guilt, anger and projection into other psycho-social domains. We should not, however, fool ourselves into thinking that 'civilization' can make us immune from primitive bloodletting practices, or that those who engage in overt sacrifice always do so out of pure callousness. Someone kills the meat we eat, making us accessories to the crime, and our civilized practices continue to create sacrificial victims and scapegoats. In many traditional cultures, belief in the efficacy of animal sacrifice is quite genuine. For example, the Hmong believe that an evil spirit (*Dab*) can steal a child's soul but that it may be fooled if the child's name is not used, or placated by the offering of an animal soul.[56] Interestingly, shamans and parents who carry out such ritual sacrifices do much to keep Hmong culture cohesive and reduce crime and delinquency. It could easily be argued that the ritual killing and eating of an animal by a cohesive community, with the goal of healing, shows greater respect for nature and the order of the organic world than the thoughtless consumption of meat from the slaughterhouse.

What does it mean that, wars notwithstanding, humankind has slowly abandoned ritual sacrifice, and that many are up in arms about animal sacrifice? Does it mean that we no longer fear the wrath of the gods or feel the need to appease them? Paradoxically, the answer is no! In many ways we feel that sacrifice and slaughter are wrong because it upsets the natural balance and is morally reprehensible. Thus, the gods that once required sacrifices now object to such practices! Did the gods become kinder and less demanding, simply because our conception of them changed? Is it all in our minds? Have we found a good argument for denouncing the reality of jealous gods or even gods in general? We can explore these questions in the final section of this chapter and consider what it is in our psychological evolution that animates the evolving mind and first sets us in relationship first to jealous, wrathful gods and then later to much kinder deities.

Into the mystic: the cosmic connection

We have already discussed 'spiritual experience' (see Chapter 3) and then touched on what Richard Bucke calls 'cosmic consciousness'.[57]

As William James indicated in his famous book title, there are many *Varieties of Religious Experience,* and among these we can include mystical experiences.[58] Putting our cards on the table from the outset, our thesis is that mystical experience is typically 'trans-rational' and therefore akin to the transitional shift from the mental to the integral consciousness structure (Table 8.1). One of its characteristics is a sense of unity or reunification with the cosmos, and, as is well known, the term 'religion' literally means to 're-connect' or 're-link' (*re-ligere*). As done throughout this book, we wish to de-emphasize the mental consciousness habit of measuring or considering some consciousness structures or states as 'raised', 'expanded' or 'higher' than others, even though this adjective is commonly used, and will be used again in trying to explain intensified states of consciousness. Thus, in our scheme (Table 8.1) we do not necessarily imply that mystical experiences confer fully integral consciousness structure, even though they may in some cases. However, we do note that many mystical experiences indicate spontaneous eruptions from the ever-present origin, even if sometimes induced by dedicated, consciousness-intensifying practices.

The term 'mysticism' comes from the Greek μυω, meaning 'to conceal', and is common to all spiritual-religious traditions, belonging to neither East nor West. This universal characteristic makes mysticism a particularly intriguing psychological phenomenon for students of consciousness. Mystical experience can be described as the subjective experience of union with the divine, with light, love, 'Godhead', 'Holy Spirit' or the *mysterium tremendum.* Thus, it usually involves awe, wonder, and a sense of the numinous and ineffable. Biographies and explanations of mystical figures by others, especially rationalists, are quite different from the subjective and experiential autobiographies of the mystics themselves.[59] Some early autobiographies use the third person rather than 'I'. Moreover, when the Ego is dissolved, the I-thou, or self-cosmos relationship is fundamentally changed. Despite the added reluctance to divulge intense, personal experience, much authentic literature and scholarship exists on mystical experience.

Before sampling historical and experiential examples let us consider a case of misunderstanding by an ultra-rationalist trying to understand a mystical message. The British biologist Peter Medawar claimed he could not follow Teilhard de Chardin's poetic and mystical discourses on evolution without a feeling of suffocation. Quite apart from the fact that Teilhard's work had been translated from the French, both were using very different languages. Ironically they both wrote books with the same title — *The Future of Man* — but entirely different messages: Teilhard's being spiritual and Medawar's being technological.[60] Medawar demanded a rational language and levelled the common, anti-mystic criticism that Teilhard's more integral expression was 'woolly' and vague. Today such criticisms are less of an obstacle to serious study of mysticism, and a growing number of researchers are keen to ask, *'What does mysticism have to teach us about consciousness?'* and how may it give us insight into scientific reality.[61] Indeed, mysticism is increasingly seen as one manifestation of the emergent, integral consciousness structure.[62]

In western mystic-spiritual traditions, Plato (428–348 BC), technically a pre-Christian pagan, used experiential mysticism as an exploration of consciousness rather than as the observance of any religious tradition. He held that the soul once shared a previous existence with the true 'Being' with which it manifests an instinctive longing to reunite. Thus, true knowledge required insight into this higher realm. This view, which was not original with Plato, has profoundly influenced western religion and there are obvious parallels with Abrahamic traditions which speak of a divine God, 'up there'. Likewise, concepts of a universal Godhead with which one may reunite, to live in a world of pure spirit, is very much at the heart of Hinduism and Buddhism giving us the concepts of *Brahman* and *Atman* noted previously. Given the profound influence Plato and other Greek philosophers, especially Aristotle, have had on the 'rational' western mind, it is somewhat ironic that Plato's 'Dialogues' had little to do with explanatory reason *(logos)*. Rather they were about direct mystical experience. Rupert Lodge points out that Plato's 'philosophy' has no prescriptive formula. There is nothing, nothing whatever, which you might conceivably discover, write down, and pass around in a printed book which could be set upon library shelves and put into the hands of young students. It is like poetry or music. You have to experience it directly, in and for yourself'.[63] Likewise Rudolf Steiner noted that: 'It is evident how

Plato feels himself in harmony with the Mysteries! He only thinks he is on the right path when it is taking him where the Mystic is to be led'.[64] This mystical place has to do with the uniquely human domain of consciousness. In *An Essay on Man* our recently introduced friend Ernst Cassirer notes that:

> We cannot discover the nature of man in the same way that
> we can detect the nature of physical things. Physical things
> may be described in terms of their objective properties,
> but man may be described and defined only in terms of his
> consciousness.[65]

A half millennium later we encounter St John's gospel, held by many to be 'the charter of Christian mysticism', and quite unlike the three other similar, or 'synoptic', gospels of Mathew, Mark and Luke. John's message is highly allegorical and devoted to the spiritual journey. It holds that humankind is 'born from above' thus bringing the Godhead — the pure being — within the reach of intelligent devotion. John the Baptist literally sees 'the Spirit descend like a dove from the sky ... to rest on him [Jesus]'. Those who follow Jesus are described as 'Children of Light' who love truth, come into the Light, and testify to the Light. Ravi Ravindra, geologist turned religious studies scholar, sees strong parallels between eastern spiritual traditions and this his 'favourite' mystical Gospel.[66] We should not take too literally the spiritual message of John or Jesus that one must be 'born again' in order to enter the 'kingdom of heaven'. This language speaks to the transformative, mystical experience, which changes consciousness irrevocably, removing the scales from one's eyes in order to see. Such experience has little or nothing to do with those institutionalized religions which demand 'blind' acceptance of Christ or God as a wrathful authority figure.

Not 'innately' mystic like John, Saul (Paul) of Tarsus reported a life-changing 'mystical experience' on the road to Damascus. He also came to believe that we all partake in the divine nature in which we 'live and move and exist'. (Acts 17:28) For Paul, Jesus was the 'unique God man'. Contemporary (AD 80–200) Gnostic tradition (from the Greek *to know*, with the Sanskrit precursor *jnana*) relied on an 'inner' or emergent soul sense of 'knowing' the divine (not the 'external' sense of knowing a thing). The Gnostic gospel of Thomas quotes Jesus as saying that both 'received their being from the same source'.[67] This almost eastern-mystic-Hindu orientation regards the creator and the created as a unity — not as separate entities.

Similar eastern influence manifests in the works of Plotinus (AD 205–270), a so-called neo-Platonist born in Egypt before moving to Rome with the hope of establishing a community (Platonopolis) where he could practise his philosophy.

Table 8.2. A Platonist/neo-Platonist schema relating cosmic unity to human consciousness.

PRIMEVAL BEING — THE ONE

Universal, Good, Beauty, Perfection

Universal intelligence / mind

Other minds — world of ideas—self-conscious reason

Like Plato, Plotinus was a 'practical transcendentalist' well versed in mystical trance, who sought to 'cut away everything that is not the One'. We can feel his 'deep-seated passion for the Absolute' when he writes, in the third person:

> ... the soul ... ceases to be itself. It belongs to God and is one
> with Him ... nothing stirred within ... neither anger, nor
> desire, nor even reason, nor a certain intellectual perception ...
> being in an ecstasy, tranquil and alone with God, he enjoyed
> and unbreakable calm.[68]

Platonic and Aristotelian philosophy (not religion *per se)* speaks to the 'vast chain of being' that descends from the highest level of perfect unity to the realms of human consciousness (see Table 8.2). Again it is interesting to compare this view with Hindu traditions that predate the Greeks by at least five hundred, if not one thousand years (see Table 8.3).

Table 8.3. A Hindu schema relating cosmic unity to human consciousness

Brahman: the impersonal unmanifest godhead

The manifest personal god Mahavishnu (three manifestations)

Brahman- creator — Vishnu sustainer — Shiva destroyer

Avatars Demigods

Atman — the divine in all of us (Brahman component)

St Augustine (AD 354–430), the Bishop of Hippo from North Africa, integrated neo-Platonism into Christianity and spoke of beauty, God, and the soul's quest to return to God. Augustine also wrote *On Free Choice of the Will* and, as Tom Cahill says, he was the first to write about subjective feelings, thus beginning the tradition of autobiographical *Confessions* and showing the strengthening of the individual soul.[69] At this time the 'monastic-ascetic' order of the Desert Fathers included St Anthony who reportedly went live in the desert to attain the state of perfection lost by Adam and Eve. He did well, supposedly living to be 100![70] All this shows us that important figures from Axial to Biblical times were experientially convinced of a higher, transcendent or divine order.

Mysticism continued to be serious business throughout the Dark Ages (AD~ 400–1100), The literature of this period is a veritable record of conscious quests. Dionysus, a mysterious fifth century Syrian monk produced *The Divine Names* and *The Mystical Theology* as practical guides to the spiritual-mystical quest. The *via negativa* or 'apophatic' way of non-language, divine ignorance or unknowing, contrasts with the kataphatic way of affirmation (through speech). Today the related biological terms 'anabolic' and 'katabolic' are sometimes used in reference to meditation. The anabolic way 'builds up' consciousness by constant affirmation and repetition (mantras) of one's creed, whereas the katabolic way 'breaks down' consciousness by emptying the mind of all thoughts and worldly distractions. Mystical practice sought to change consciousness and effect a reunion (*religere*) with the divine realm — the *mysterium tremendum*.

Although the experiential fruits of such labours were uplifting, many avowed mystics faced stiff opposition from conservative ecclesiastical quarters. The ninth century Irish monk John Scotus Erigena translated Dionysius' text *On the division of nature* which in turn influenced ", a twelfth century University of Paris scholar, who spoke of the *Doctrine of Free Spirit*.[71] This 'Dionysian' leap catapults us from the Dark Ages to the medieval age of Renewal. Amaury's doctrine was 'officially' denounced by the council of Vienna in 1312, and Nicholas of Basle was burned as a heretic for subscribing to Amaury's belief that 'all spirits return to God'. We certainly hope that was true in his case! Likewise, the Dominican monk Meister Eckhart (1260–1327), held that 'man should not rest satisfied with an imaginary God', and as previously noted, his 'whatever you call him [God] — he is not that' could have come directly from Taoism. His equally famous statement, 'the eyes in which I see God is the same in which he sees me' smacked, in the eyes of the church, of heretical self-deification.[72] We are again reminded of Charles Péguy's quote from Chapter 4 that: 'Everything begins in mysticism and ends in politics'.

In our light-speed journey through western religious history, we see the co-evolution of age-old religious sentiments (mystical intimations of divine union and belonging to the cosmos) and the emergence of individuality and individual conscience. St Bernard of Clairvaux

(1090–1153) founder of the Knights Templar, used the metaphor of God as bridegroom and seeker as bride, a stance echoed by Ramon Llull, a thirteenth century Franciscan from Catalonia, in *The Book of the Lover and the Beloved*. God appeared incarnate so as to draw humankind's carnal love towards spiritual love, and so dawned the chivalric tradition of courtly love (Chapter 9). The eleventh through thirteenth century period of renewal and scholasticism, especially in Spain, spawned the remarkable 'Andalusian Age' reaching a zenith of Islamic tolerance that compared contrasted very favourably with the persecution of the Cathars (discussed in Chapter 3) taking place in France just over the Pyrenees. Meanwhile England and France founded their first universities. These developments were a socio-spiritual prelude to the humanistic " that flourished in fourteenth through sixteenth century Europe. Humanism, and belief in direct 'gnostic' communion with the divine, challenged Catholic dogma and eventually, in 1523, led to the Protestant Reformation.[73]

Let us supplement this historical chronicle with and experiential sense of the eruption of spiritual fire so beautifully conveyed in "'s delightful portrait of St Francis of Assisi (1282–1326). He writes: '... the whole philosophy of St Francis revolved around the idea of a new supernatural light on natural things, which meant the ultimate recovery not the ultimate refusal of natural things'. His triple vow of obedience, poverty and chastity, was no intellectual exercise, it meant embracing lepers, throwing himself into the world with fearless abandon, to re-enter and be reborn a manual labourer, rebuilding a church brick by brick. He was a troubadour, in love with poverty, appreciating the nothing of which everything is made. 'The man who went into the cave was not the man who came out again'. It was as if he had died and become a blessed spirit. 'Never was a man so little afraid of his own promises. His life was one riot of rash vows; rash vows that turned out right'. 'He would become more and more a fool; he would be the court fool of paradise.' The man who has seen the human hierarchy upside down through his vision of divine reality, may have something of the appearance of a lunatic. He was a man who had lost his name while preserving his nature, who paradoxically was transported with joy to discover that he was in infinite debt, in debt to the divine. His mysticism was 'so close to the commons sense of a child ... [He had] no difficulty ... understanding that God made the dog and the cat'.[74] Thus Francis stirred the world with his spiritual enthusiasm, knowing divine love as a reality. Popular legend has it that when he broke bread with St Clare, the people of Assisi thought that the trees and the holy house were on fire. When they approached they found the holy pair surrounded by a red halo, discussing divine love.

Is this exaggeration, or were mystics like Francis really ablaze with divine light, and baptized by fire? Is it mere poetry to describe John of the Cross (1541–91) as a 'spirit of ardent flame', like Francis, 'on fire' with the love of God, the supreme spiritual mountaineer? Is this hyperbole or an honest representation of a cultivated spiritual consciousness from a previous millennium? With the red halo legend in mind, let us turn to the reports of a few modern mystics. Our friend Richard Bucke, author of *Cosmic Consciousness* writes, in the third person of his own 1873 experience:

> It was in the early spring ... of his thirty-sixth year. He and two
> friends had spent the evening reading Wordsworth, Shelley,
> Keats, Browning and especially Whitman. They parted ...
> his mind deeply under the influence of the ideas, images and
> emotions called up by the reading ... All at once, without
> warning of any kind, he found himself wrapped around as if it
> were by a flame coloured cloud ... the next, he knew that the
> light was within himself. Directly afterwards came upon him
> a sense of exultation, of immense joyousness accompanied
> or immediately followed by an intellectual illumination
> quite impossible to describe. Into his brain streamed one
> momentary lightning flash of Brahmic Splendour which has
> ever since lightened his life; upon his heart fell one drop of
> Brahmic Bliss, leaving thenceforward for always an aftertaste
> of heaven ... he learned more within a few seconds during
> which the illumination lasted than in previous months or
> even years of study.[75]

Of his experience, Gopi Krishna writes:

> Suddenly, with a roar like that of a waterfall, I felt a stream of
> golden liquid entering my brain through the spinal cord. The
> illumination grew brighter and brighter, the roaring louder
> and louder ... I [became] as vast circle of consciousness in
> which the body was but a point, bathed in light and in a state
> of exaltation and happiness impossible to describe.[76]

Neither Bucke nor Krishna speak of their experiences in the context of an institutionalized religion. Moreover, like the following example, they describe what is more than a mere psychological experience. The physical and emotional sensations also suggest a biological experience. Krishna's experience is what is known as a kundalini awakening or kundalini release (see also Chapter 3 above). In recent years the Religious Experiences Research Centre, founded by Alister Hardy, now at the University of Wales, has collected reports of many similar experiences in which the subjects recognize the kundalini phenomenon. The following is representative:

> While ... in a deep state of concentration ... I experienced a
> lot of physical sensations like electricity/tingling over my
> forehead and over the sides of my head and body.. at the top
> of my vision came a bright white sparkling ball made up of
> sliver white rays/blades of light flickering ... very clear [but
> with eyes closed]. The physical sensations intensified over
> my whole body to such a degree that I felt numb ... and the
> electricity was intense. .. At this moment I had a completely
> overwhelming sense of bliss/love. It was awesome — it made
> everything else I had ever experienced pale into comparison.
> Then a few minutes later, a massive surge of upwards energy
> towards this light actually made me stand up from a sitting
> meditation as I felt that I was being shot up out of my body ...
> I believe that I experienced a kundalini awakening where the
> kundalini energy shoots upwards toward the crown chakra.
> An amazing experience.[77]

Could it be that such intense experience, with its physical, emotional, psychological and spiritual components is a normal biological phenomenon, associated with the evolution of the nervous system, or the energetic and psychological evolution of the whole human species? After all we are physically, emotionally, psychologically (and presumably spiritually) different from our primate ancestors. So why should evolution not continue to make us psychologically different from our earliest human forebears? Before we explore this possibility in our final chapter, we should note that individual spiritual experience is ongoing in human history. According to scholars like Alister Hardy, such bio-psychological experiences are increasingly reported in language that lacks traditional religious vocabulary. Nevertheless the spiritual intensity is felt no less deeply than it was by the most devoted medieval mystic. Many words and phrases in the contemporary mystic's lexicon are the same as in yesteryear: 'illumination', 'sense of divine presence', 'baptism by fire' and 'peace that passeth all understanding'. Such experience, which clearly affects the body's energy field, almost certainly has biochemical 'correlates' with dopamine and other brain chemicals, but such materialist explanations are only part of the story (not least because, among those adept at meditation, they are the effect as much as the cause). Chemicals in the brain may bring about a sense of wellbeing, but what of the extraordinary vital energy that can literally launch one off the ground? What of the long-term effects on consciousness reported by almost all individuals who have had such experiences? Why do so many subjects insist that they have experienced what Krishna calls the species' 'evolutionary energy, ' and why do they sense, if they are not traditionally religious, such a deeply meaningful sense of divine blessing? Are humans fundamentally spiritual beings, as so many gurus remind us? Whichever way we view it, as the twenty first century dawns our psycho-spiritual makeup is as much a part of the evolving human experience as it ever has been.

9. Closing the Circle

The poetic impulse: its psycho-spiritual origins

Throughout the book's early chapters, we attempted to build a case for the understanding of human consciousness through the integration of many fields of knowledge, among them psychology, biology, cultural history and evolutionary theory. This follows Gebser's thesis that cultural artifacts are the manifestation of consciousness. We also follow Jung's reasoning that consciousness is a necessary prerequisite for representing the world to ourselves, and for conceptualizing any such intellectual fields in the first place using reason *(logos)* and all the 'ologies' it creates. As we proceeded, we have also argued that consciousness changes dynamically and that new properties, or consciousness structures 'emerge' spontaneously, or as the result of intrinsic developmental dynamics that affect our biological and psychological make-up. This view was particularly emphasized in the previous chapter where we argued that the emergence of language and a religious or spiritual sensibility was the inevitable result of becoming self-conscious. Our ongoing discussion of such novel, emergent faculties as spiritual or mystical experience and intellectual and artistic genius, suggests that becoming self-conscious is just one shift in consciousness structure that individuals undergo as Ego-consciousness intensifies and we 'separate' individually and collectively, from the universal ground of unconscious nature.

'Beyond' Ego-consciousness, (and here spatial metaphors are inadequate), the realm of cosmic consciousness or the 'higher' faculties of the Superego are, as Gebser says, ever-present. As every mystical, spiritual experience informs us, transition or transmutation across this threshold, however subtle or abrupt, leaves the individual with a profound sense of reunification with the fundamental, universal ground of being. It is for this reason that, almost without exception, mystics interpret their experiences in remarkably consistent ways. First, they conclude that cosmic consciousness is a natural and desirable evolutionary development that enhances intellectual and moral sensibilities, softening Ego-consciousness and melting it back into the universal ground. In many cases, if the shift is marked enough, individuals may develop new supersensory faculties such as clairvoyance, telepathy or healing powers, as if tuning into new consciousness frequencies. Second, and related to the first experiences, they probably conclude that consciousness is the primary datum and not some epiphenomenon or side-effect of neural activity.[1] The latter view only seems possible where analytical mental-rational states predominate. Thirdly, the self-realized mystic often claims to 'know' they are immortal and so believes that death is not the end but merely a 'change of state, ' and a chance for rebirth.

Many of the mystics mentioned in Chapter 8, although now regarded as enlightened holy men and women (illuminati) were regarded in their day as dangerous, even subversive, renegades. Although many may have been 'devoted' monks or nuns, while others were simple lay persons, they were all, after the experiences they reported, far from being slavish disciples of institutionalized religious doctrine and dogma. On the contrary, their experiences created such profound shifts in conscience that many were compelled to challenge institutionalized religious doctrine, often at great personal risk. In addition to those like Eckhart who died for their heresies, even twentieth century mystics like Teilhard de Chardin, were severely punished by the Church. All Teilhard did was write provocative evolutionary philosophy that attempted to reconcile science and spirituality and maintain the Church's relevance in an age of empirical scepticism. For this he was prevented from publishing during his lifetime and exiled as far from Europe as possible.[2] Amaury of Bene, whom we met in the last chapter, also said the entire Church would pass away and in place would emerge a new era of human spirit based on inner consciousness of God.

To avoid appearing western- or Christian-centric, and to stress the point that mystics often occupy a paradoxical position as both spiritually enlightened devotees and religious heretics, let us continue our study of mysticism by taking an eastward journey through the Middle East and on to India and the Far East. On the way, let us also bear in mind that our aim is to explore the psychological 'truth value' that appears to be inherent in the mystic, cosmic consciousness adventure. We stress that mysticism by its profoundly psychological characteristics, and novel, even radical emergent properties generally has little to do with exclusive institutionalized religious doctrines. On the contrary, it appears more like a bridge between the best inclusive spiritual traditions and compassionate humanism.

Until recently western culture knew little about Islam, and probably even less about Islamic mysticism. Unfortunately, much of what it has learned in recent years has pertained to exclusive fundamentalist sects that have been active in the political and combat arenas. But there is a whole other domain of Islamic spirituality, popularly known as Sufism (which derives from the word *souf*, and refers to the coarse woollen garments worn by the Sufis).[3] Sufism may have been influenced by Christian mysticism, various gnostic traditions, Asian shamanism and Indian yogis.[4] Popularly known as whirling dervishes for their ritual dancing that raised consciousness, Sufis have also been referred to as 'intoxicated'. This clearly has nothing to do with alcohol, but instead refers to 'ecstatic' or trance-like states more akin to shamanistic practices. In the Sufi tradition, the idea of complete union (unicity or *tawhid*) with God reflects a type of pantheism which might be construed as a purer form of mysticism than the Christian variety and more like the Vedantist tradition.

The list of famous Moslem mystics is at least as long and fascinating as the list of western Europeans given in Chapter 8.[5] It is also interesting to note how, here again, eccentric behaviour was often regarded as heretical and frequently severely punished as a result. The eighth century mystic Rabi'ah al-Adawiyah from Basra, Iraq, ran through the streets with a torch in one hand, to light the way to heaven, and with water in the other, to quench the fires of hell. She is described as having been on fire with love and longing, and having taught that love alone was the only guide on the mystic path. Ziyad B. al-Arabi (ninth century) spoke of indefinable ecstasy as the vision of the heart, known only through experience. Though the vision ends, the knowledge remains. Abu Yazid al-Bistami (d. 875) said God was in his soul and declared: 'Glory to me. How great is my majesty.' He held that to know God was to know 'the Great Silence' and experience the 'dissolution of individuality'. In a similar vein Al Hallaj (857–922) stated, just as Jesus had done, 'I am the truth'. He was crucified and burned for his troubles.

Ibn Sina, also known as Avicenna (d. 1037), spoke specifically of the stages of mysticism:

1. knowing the way;
2. self-discipline, removing all but God and conscience from attention so as to be attracted to higher things;
3. loss of sensual desires, flashes of divine light appear;
4. seeing God in all things;
5. becoming accustomed to God's presence and seeing divine light constantly as a divine flame;
6. contemplating God in himself, until his inmost soul becomes a perfect mirror, and he has complete union with God.

Abu Hamid al-Ghazzali, a divinity professor in Baghdad in 1091, confessed that, for him, theory was easier than practice. Thus, some say he was not a true mystic, but more equivalent to Augustine or Thomas Aquinas in Christianity. This is a perhaps a good example of intellectually-honest scholasticism that recognizes the difference between theory and practice.

Ibn al-Arabi (1165–1240) was said to be one of the greatest Sufis. He held that nothing exists except Allah: that is, everything signifies God. Man is God and God is man. This very Hindu sentiment is perhaps not surprising given the cultural contact between Islam and India. He also spoke of the 'Beatific Vision' that manifests as epiphany to the 'elect' or chosen few here on Earth. Holding, on the one hand, that there is but one true epiphany may seem dogmatic to modern sensibilities, but the qualifier, that there are many manifestations because of the differences among those who receive it, is decidedly pluralistic and postmodern, and could easily have come from the pen of a modern student of spiritual experience. Ibn al-Arabi certainly kept his pen busy. He wrote the longest known poem (180,000 lines) and said that if all the trees were pens and all the seas ink, and they began writing, they could not possibly exhaust all the names which God has given to his creations. This again echoes various Hindu and Buddhist notions (of 33,000 or 10,000 things). He evidently proved his point by never running out of ink!

Last on our present list is Jalal-Din Rumi (d. 1273), ironically today the best-selling poet in the USA, who wrote classic poems entitled *Unity of Spirit, The One True Light* and *The Mystic Way*. In the first poem he writes:

> 'Tis wrong to think that the vicar
> and He whom the vicar represents
> are two.
> To the form-worshipper they are two;
> when you have escaped from consciousness of form,
> they are one.
> Whilst you regard the form you are seeing double ...
> In things spiritual there is no partition, no number, no
> individuals ...
> Mortify rebellious form till it wastes away: unearth the
> treasure of Unity.[6]

This don't-believe-in-form philosophy is a type of quantum physics insight which also has obvious predecessors in Hinduism: that is, the sense perceptible world is *maya* (illusion). The 'unearth — unity' quest is also obviously parallel to the physicist's quest for Grand Unifying Theories (GUTs). This forces us to ask again whether contemporary science is not actually pursuing the same quest for unity that mystics have been seeking all along. The philosopher Michael Polanyi makes a compelling point when noting that scientists believe in an 'ever continuing possibility of revealing still hidden truth'.[7] This amounts to a 'belief in a spiritual reality' or a 'perception of the ... hidden reality in nature'.[8]

In *The One True Light,* Rumi promulgates a famous message of unity underlying the differences so prevalent in different belief systems, saying: 'O thou who art the kernel of existence, the disagreement between Moslem, Zoroastrian and Jew depends on the standpoint.' He goes on with the story of the elephant being examined in the darkness by scholars of different backgrounds. One felt its trunk, saying it was like a water pipe, another touched its ear and said it was like a fan, yet another touched its leg and declared it was like a pillar, while yet another felt its back and said it was like a throne. Rumi concludes: 'Had each of them held a lighted candle, there would have been no contradiction in their words'.[9]

Such sentiments are of course open to all kinds of rational, discursive and explanatory analysis. The sociologists and historians will say, here is a man advocating religious tolerance and a pluralistic worldview. The scientist may remark on his recognition of the intangibility of matter and ability to state poetically what the quantum physicist is inclined to say with equations. The religious studies scholar will reiterate that there is no significant difference between mystics the world over. Finally, the philosopher recognizes a person with considerable insight and something interesting to say about the nature of reality.

Paradoxically, the mystic may describe the mystical experience itself as beyond words, then be inspired to write hundreds of thousands of poetic lines reflecting high enthusiasm for the new insights gained. Here two analytical observations seem pertinent. If indeed, our ancestors lacked true human language before they became self-conscious and emerged from the non-dualistic, unconscious state, is there not a certain logic in recognizing that the dissolution of the self-conscious Ego-state (mental structure) might also involve a linguistic shift in its 'return' journey into universal or cosmic consciousness. Full conscious immersion (re-immersion) in the universal ground of consciousness allows for clairvoyance and may dispense with the need for language. Thompson infers that: 'searching in the past with language for the origin of language, one is approaching an edge ... very similar to the one ... when one moves out of linguistic thinking in meditation'.[10] Similarly it seems logical to note that just as our early language was poetic, reflecting our emotional response to nature and the cosmos, so too we necessarily rediscover our poetic expression as we experience re-entry into the 'trans-rational' universal ground. As the following 'timeline' schema (see Table 9.1) illustrates, discursive, analytical and dualistic (subject-object) language is sandwiched between our poetic origins and our poetic 'future'. This is consistent with a number of lines of evidence. Most obviously, for example, is the fact that different consciousness structures, generate different linguistic meanings, and as we have noted on more than one occasion, one cannot solve new problems with the same consciousness (or the same language) that created them. Barfield, the great genius in this field, demonstrated this only too well when he showed that the pre-rational language of mythic 'participation' was quite different, indeed sometimes opposite from, that of the abstract mental structure that promulgates dualistic subject-object 'separation'. But as consciousness intensifies, 'separation' gives way to Barfield's 'final participation' and the need for a new language structure.[11] So, while the new quantum physics proves that the observer cannot objectively divorce him/herself from the observed, so the mystic reports that Ego-separation is an illusion and that we are indeed conscious participants in the universal ground of consciousness. The new physics seems so intelligently complex and mysterious that it is apt to send many cosmologists, including Einstein into raptures of awe about the wonders of the universe and the mind of God.[12] Such responses seem absolutely identical to those of the mystics contemplating the same universe. In this we are reminded of how the new physics paradigm is almost a century ahead of much biology which is often still stuck in the mechanical-biochemical paradigm, and less susceptible to the 'awe' that comes from contemplating subtle energies, biophotons and the like.[13] As a result Rupert Sheldrake, himself an accomplished biologist, made the wry comment that after Darwinism (especially twentieth century neo-Darwinism) God was kicked out of the front door of the biology department, but went round to the back door of the physics department.[14]

When Nietzsche said, famously, that 'God is dead', he meant that rationalism had done away with the need for institutionalized religion in many intellectual circles.[15] He was a brilliant man and he used the God metaphor thoughtfully. However, he could not foresee the equally famous anonymous graffiti response: 'Nietzsche is Dead'. Nor was he speaking for awe-struck cosmologists and mystics. Nietzsche's metaphor is reiterated by the likes of Richard Dawkins who claims that the theory of evolution allows one to be an 'intellectually fulfilled atheist'.[16] But such intellectually based analyses do not fully address all dimensions of knowledge, intuition and consciousness nor do they take account of the 'novel' emotional and spiritual impulses that arise whenever we discover new domains of consciousness that enable us to integrate more fully with the cosmic mystery. Remember that the intellectual-emotional (head-heart) pendulum keeps swinging (see Chapter 3), and falling in love with the cosmos all over again is always inclined to bring out the poet in us. We let Rumi have the last word in this section:

The Mystic Way

Our speech and action is the outer journey
Our inner journey is above the sky
The body travels on its dusty way
The spirit walks, like Jesus, on the sea..

Jalal-Din. Rumi[17]

Table 9.1 Linguistic evolution as a corollary of changing consciousness structures.

Pre-linguistic/ unconscious	Early mythic- poetic phase	Rational/ discursive	Mystic-poetic phase	Post-linguistic/ intuitive

Original participation		Separation	Final participation	
Simple consciousness		Self-consciousness	Cosmic consciousness	

Mystics and scientists: the same universal truths

Harry Hunt poses the question: 'Might mystics intuit some of what physicists calculate?'[18] His answer is affirmative. He cites the case of the famous physicist Niels Bohr who confessed to reading William James and getting the idea of 'quantum complementarity' and indeterminism from James' ideas about the alternate continuous flow (stream) and pulses of consciousness. So introspective psychology helped illuminate physical reality. Conversely, in contemporary consciousness studies, the 'Ambient Ecological Array of Perception' model borrows from fluid dynamics (and indirectly from James and evolutionary theory) to explain that we (all mobile animals) are dynamic observers flowing through an environment that changes constantly.[19]

Hunt argues that mystics, like physicists and other profound philosophical thinkers are able to feel or intuit the cosmos as a living presence, like consciousness itself. These intuitions have what Hunt calls an authentic 'truth value'. For example, physicists and mystics both agree that linear time is an illusion. Other shared notions include intimations about 'complexity' which intuit the emergence of higher order or organization (consciousness) being latent in systems. Thus 'inwardness' must be latent in matter if consciousness is not an accident in an otherwise alien universe. This was exactly the position of Teilhard de Chardin in holding that all matter had degrees of interiorness.[20] Complex patterns like mandalas and psychedelic visions mirror nature's complexity, and were the object of much study by Jung, because of their importance to some of his patients.[21] If organization principles in the physical universe and conscious patterns of perception are similar/resonant, it is a commonsense conclusion that they reflect the same reality and that metaphors for consciousness and physics will converge in concepts of unity. This principle echoes Spinoza's proposition that 'the order and connection of ideas is the same as the order and connection of things'.[22]

There is little reason to doubt that intelligent and sensitive scientists, intellectuals and mystics, are capable of perceiving patterns of cosmic organization that have what the rational and intuitive mind calls an 'elegance' and an 'internal consistency' that makes for compelling theories and explanations. If this makes the scientists versus mystics dualism an unnecessary distinction, a better approach is to consider that we are all together in this search for knowledge, intensified consciousness and deeper universal understanding. Indeed there are some organizations like the Scientific and Medical Network that hold regular 'Scientists and Mystics' conferences.[23] If the truth value of some valuable insight is verified, it is perhaps of secondary importance as to exactly how it was obtained. This question of how insight or illumination is obtained is tricky, and may even be forgotten in the excitement of the experience. In the case of intuiting and describing reality as a field, with nonlocal quantum properties and so forth, a mystic may obtain insight through meditation, whereas a physicist or cosmologist may, thanks to a gift for mathematics, thrash out an understanding though intense intellectual effort. The difficulty of trying to understanding some inferred underlying cause may be a clue to the fact that we are asking the wrong questions. Why does a mystic meditate? Why does another have a gift for mathematics? We may have to accept that in a universe of consciousness, where information is ubiquitous, tuning into new channels that provide new insights is to be expected. Thompson provides an interesting example of two different approaches arriving at similar conclusions about the important role sexual reproduction played in the acceleration of evolution. This thesis is well known to students of paleontology and evolution, and Thompson cites the work of our friend E.O. Wilson on sociobiology as a representative example of a scientific narrative stating that, 'entire populations evolve faster when they reproduce by sex'.[24] This is compared with the 'automatic writing' record of a 'deep-trance medium' who speaks, as if straight from the Gnostic creation myths, of a divine 'experiment for the wellbeing of earth and its evolution' to speed up evolution once they had 'seen that the process ... was not so quick'.[25]

Many scientists might recoil at the thought that useful empirical information could be obtained from a spiritual source by a deep-trance medium, and one can easily imagine the hasty explanations designed to discount the report: 'He'd obviously read a book on evolution sometime previously'. Such speculative explanations, beside being difficult to prove, merely reinforce the empiricists belief in the superiority of their method, while discounting the truth value of the result. Having said this, many scientific explanations of the nature of reality, especially the quantum domain, and the mysterious origins of the universe, life and our own species, are so perplexing and beyond full and unambiguous rational explanation that scientists themselves resort to the metaphor of 'storytelling' or narrative, and may even use the term 'myth'.[26] Thompson states that:

> When the scientist is ignorant of myth, cosmology, and literature, the quality of his narrative can be simplistic and naive ... but when the scientist is gifted with an imagination and artistic ability ... the narrative takes on the best qualities of storytelling [and] begins to move beneath the shallow stream of conventional empiricism to express the deeper dimension of consciousness.[27]

This amounts to saying that the scientist is better off being broadly-educated and open-minded rather than narrowly trained and dogmatic. Exactly the same concerns preoccupied the enlightened mystics when faced with the narrow and naive dogmas of intractable religious institutions.

Crowning glories: kundalini shock and awe

We saw in Chapter 3 how spiritual, or mystical experience can overtake the individual rapidly, and with surprising long term results which significantly enhance what Richard Bucke called the intellectual and moral sense, and what Thompson refers to as the deeper dimension of consciousness. Although the experience may involve being overcome by an 'oceanic feeling'[28] as if the boundaries of the body were expanding and melting into the surrounding universe, the experience may involve much stronger sensations. Testimony given in Chapter 8 indicates the oceanic feeling may be accompanied by a perceptible warm glow or even dazzling, blinding light. In some cases, however, the experience involves an 'electric' release of nervous energy in the spinal column which can be a severe 'shock' to the system, as well as powerful testimony to the latent vital energy in the body. In almost all such cases the effect on consciousness is to inspire awe and various degrees of conviction that the experience has promoted a positive shift in consciousness to a higher, deeper or more intensified level. Even when such heightened or ecstatic feelings fade with time they are never forgotten. It is as if the body-mind, having once experienced a new frequency, continues to resonate on that level.

Probably the best example of mind-body integration in the aspiration for higher or intensified consciousness is seen in kundalini yoga, which explicitly aims to channel pranic energy from the lower centres (chakras) to the higher centres, by balancing controlled physical and mental (meditative) practice. As previously noted the term chakra means 'wheel' and refers to nexuses or vortices of energy, associated, in ascending order with the basal spine (anal chakra), genital, navel, heart, lower throat, eyebrow and crown (see Figure. 9.1) which each carry their 'own psychological significance'.[29] These centres are deeply connected with the phenomenon of the rising kundalini, a posterior-anterior flow of energy, deliberately cultivated by some yogis, but sometimes released spontaneously.[30] Each chakra has a respective colour — red, orange, yellow, green, blue, purple and violet — which together correspond to the ascending frequency of the light spectrum.

Yoga simply means 'yoke' and refers to the practice of linking one's spirit to God or the divine. Patanjali's *Yoga Sutras* (sutra means 'book', teaching or philosophy) probably date from between 100 BC and AD 300 although their antecedents have long and complex histories. Some seals showing figures seated in the lotus position suggest yoga's very early roots in the Harrapan Culture of the Indus valley which dates back to around 2700–2800 BC, a time coincident with the very earliest record we have of what was to become Hindu culture. Patanjali only codified or systematized the physical, mental, moral and meditative practices already extant, and probably well established at the time of the Upanishad Ferment around 800–500 BC.[31] This formalization helped

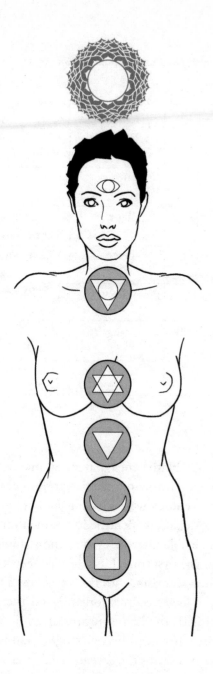

Figure 9.1. The seven chakras.

Table 9.2. The hierarchy of chakras in kundalini yoga.

Chakra name (number)	Colour	Associations
Sahasrara crown (7)	Ultraviolet	Transcendental God consciousness
Ajna forehead (6)	purple	Spiritual third eye; subtle body
Visuddha throat (5)	blue	Artistic expression; consciousness of eternity
Anahata solar plexus (4)	green	Conscious human being; Heart chakra
Malipura navel (3)	yellow	Passions; emotions
Savdhisthana genitals (2)	orange	Desire
Muladhara anal (1)	red	Earth

give us variations like Hatha yoga, a physical practice now familiar in the west, Yama yoga , involving pacifist philosophy, and tantric and kundalini yoga with consciousness raising connotations.[32]

The hierarchy of chakras identified in kundalini yoga (Table 9.2) should be of considerable interest, even to the analytical western mind encountering it for the first time. First the colour spectrum is exactly as we see it in a rainbow or prism, with the warm, lower frequency colours associated with the lower centres (organs) and the higher frequency colours associated with the higher organs. Tradition says that: 'when kundalini-sakti rises, her head becomes light'[33] and clearly it equates illumination or activation of the seventh chakra as a 'crowning glory' among consciousness experiences. The pure 'white' light metaphor is astonishingly pervasive in all human culture. It appears repeatedly, not just in mystical and spiritual texts, but in physics, too, where light (the speed of light) symbolizes the ultimate law of the universe, a threshold beyond which we cannot go. To evolve to a higher level is to become enlightened. Again Gebser reminds us that it is the eye that is the symbol of mental culture as we speak of insight and shedding light on a subject. But we may ponder, did the ancients, living in dark caves and pondering the shadows on the wall, revere light in the same way as we do? Two authors, Max Müller[34] and Richard Bucke[35] have suggested that the ancients may have perceived only the low frequency end of the light spectrum. The evidence for this conjecture comes from the lack of words for blue and the high frequency end of the spectrum. Philip Snow (son of the famous 'Two Cultures' C.P. Snow) hints at a similar linguistic situation in medieval east Africa (Ethiopia) where there is also no word for blue.[36] It is certainly true that many ancient cave paintings are dominated by blacks, browns, oranges and reds (see book cover). There may be other possible explanations for this, such as the absence of blue pigments in some areas, but this would raise other questions such as how colour perception and consciousness is influenced by environment. Clearly the chromatic world of the Eskimo and the African differ considerably.

In making the suggestion that ancient cultures may not have perceived what we call the higher frequency colours, we should stress that we mean only that they may not have perceived them in the same way. In an interesting parallel to the Hindu perception of the seven chakras, the Australian aborigines had a concept of seven levels of vibration (see Figure 9.2) in the 'Dowie', a subtle level of consciousness often associated with the afterlife or astral plane.[37] The number seven comes up frequently in relation to levels of consciousness. For example St Teresa of Avila wrote of the 'seven mansions' of 'the interior castle' — a theme reiterated in Thomas Merton's Seven Story Mansion. Likewise, the relatively uneducated mystic Jacob Boehme (1575–1625) perceived seven properties (qualities of the universe which unfolded out of God) of which the Seventh Property became perfect union, Paradise and 'eternal nature'.[38] Whether any of these mystics were directly or indirectly influenced by oriental thought or not, is an interesting matter for conjecture.

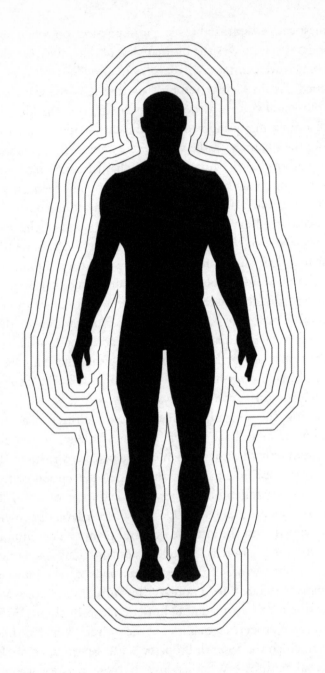

Figure 9.2. A depiction of seven levels of vibration in the Australian aboriginal subtle body, the 'Dowie'. (After Havecker 1987.)

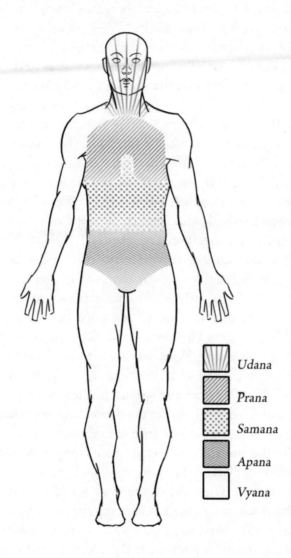

Figure 9.3. The pranic body. (After Saraswati 2002.)

Although chakras have attracted the attention of western researchers like Carl Jung, the subject remains somewhat taboo in academic circles and is treated overcautiously, leading to the suggestion that we cannot find the biological basis of the phenomenon.[39] We suggest that this is no longer the case. Work on the molecular and electromagnetic coherence of the body, not to mention stored light energy (biophotons), as discussed in Chapter 7, is providing confirmation that the body can be known as a field just as ancient and modern yogis have reported from their direct experience.

In India's Vedic tradition, dating back to around 1500 BC, the concept of *prana* or essential life force, like the Chinese concept of chi *(qi)* cannot be separated from the concept of the chakras.[40] In a parallel with the Schadian schema described in Chapter 6, the pranic body is seen as having five elements, which in ascending order from posterior to anterior are: Earth (elimination, reproduction and mechanical energy); Water (belly and digestion and chemical energy); Fire (respiration, circulation and thermal energy); Air (brain, senses and electrical energy); and Space (coordination of other elements and nuclear energy).[41] (See Figure 9.3.) Unfortunately, modern western science tends to regard the metaphors of earth, water, fire, air and space, as archaic and unscientific relics of bygone ages. However, as the Schadian schema shows, there is nothing unscientific about the association of tangible matter and fluid (earth and water) with the lower organs of the body that control digestion, elimination, reproduction and circulation. Similarly, fire and air are associated with thermoregulation and respiration, which are both physiological processes that are profoundly integrated. Finally, there is no dispute that our dominant long range senses (vision, hearing and smell) take place in the realm of air and space.

We can effect a fruitful marriage of Schadian thinking with the pranic body schema by again reiterating the point made in Chapter 6 that during the course of evolution the vertebrates have progressively internalized organs beginning at the anterior (brain) and eventually leading to the posterior reproductive system (presented here again for emphasis as Table 9.3). These progressively internalized organs (brain, lungs, circulation, thermoregulation and reproductive system) correspond broadly to the space, air, fire, water and earth elements of the pranic system familiar to yogic traditions.

Table 9.3. *The anterior to posterior internalization of organs during vertebrate evolution.*

Fish	Amphibian	Reptile	Bird	Mammal
Central nervous system	Respiratory system	Fluid system	Thermo-regulation	Reproduction
Brain	Lungs	Heart	Viscera	Uterus

In the yogic tradition it is possible, through careful practice, to distribute higher consciousness through the whole body-mind so that normally unconscious processes of digestion and heat regulation can be controlled. Such conscious control is also practised to a lesser degree in most western traditions, where various forms of physical, emotional, intellectual self-control (learning) and enlightenment are as much prized as the higher faculties aspired to in yogic practice. Here we can note that some of the most thorough, well informed and up-to-date physiological science shows us that the distribution of senses throughout the body does in fact range across a spectrum from highly conscious and open to the outside world (cosmos) to closed and very unconscious (Figure 9.4).[42] Although, as Jung noted there are significant cultural differences between East and West that arise from very different histories, mythologies and worldviews, there is no doubt that, in recent generations, knowledge of kundalini yoga has spread in western culture, and that the incidence of the experience is slowly coming to light as understanding of spontaneous, or meditation-induced spiritual experience becomes more widespread (see Chapter 8).[43]

If more evidence is needed to support the idea that the body-mind is an energetic field of dynamic currents which affect the flux of consciousness and health, we can turn to Chinese medicine and examine the *chi* carrying meridians associated with the major organs

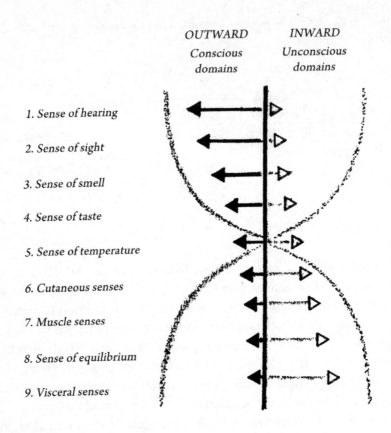

Figure 9.4. The distribution of higher (conscious) and lower unconscious senses in the human body. (After Rohen 2007.)

(gut, heart, lungs, kidneys, small and large intestine, and so on) and the ancient practice of acupuncture. The term 'meridian' is the western (French) translation of *Jing-luo* meaning a thread that runs through or connects in the form of a net or web. Dramatic evidence of the antiquity of knowledge concerning energy meridians was revealed by the discovery of a Bronze Age traveler murdered in the European Alps some 5,300 years ago. This individual nicknamed Ötzi, or the 'Iceman' had numerous tattoos associated with acupuncture meridians.[44] There

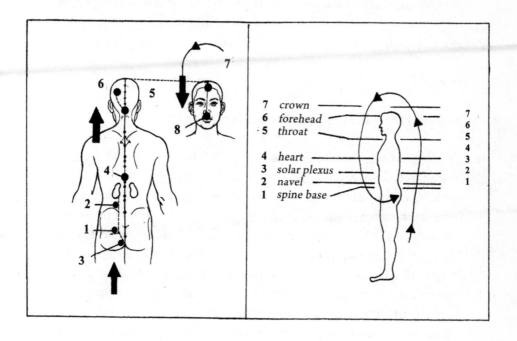

Figure 9.5. A comparison of the Governing Vessel meridian from acupuncture (after Kaptchuk 1983) and the path of Kundalini energy (after Bentov 1977). Note that in both cases, rising energy also descends.

are twelve regular (five Yin, six Yang and one independent) and eight extra meridians, of which two — the Governing Vessel and the Conception Vessel — are considered major meridians because they have many points that are independent of the other meridians.[45]

The Governing Vessel begins in the pelvic cavity and follows the middle of the spinal column up into the brain, and continues over the crown of the head and back down to the forehead, nose and upper gum (see Figure 9.5). It is interesting to note that this meridian reaches the brain at point six and that as well as ascending, there is a final phase of descent to the nose and the upper lip. In the yogic tradition the rising of the kundalini can be dangerous to the body's nervous system, especially if the current of energy does not descend and complete the circuit.[46] What goes up must come down! This is the essence of the life cycle.

The courtship between science and spirituality

Throughout this book we have been aware of the need to tread carefully in our use of such terms as science and spirituality, both of which ultimately have very broad meanings. Moreover, both are phenomena inseparable from cultural history and so share common sources and complex relationships in the domain of human consciousness. However, in an integral paradigm where dualism is acknowledged, but consciously de-emphasized, there should not, by definition, be a wide 'two cultures' gulf between these domains. In our optimistic opinion, it is an encouraging sign of a collective movement towards a more integral paradigm that the possibility of a marriage between science-spirituality (or science and religion) today receives so much serious press. As globalization forces different cultures into close proximity, so different spiritual, religious and secular norms and ethics find themselves in direct communication. The challenges faced in integrating different cultural and spiritual traditions have been discussed by many commentators, as we did in our introduction to Chapter 8. The results have varied greatly, and it is clear that while well behaved scientists and mystics often enjoy a comfortable meeting of the minds, in other cases the two cultures are unable to engage in constructive dialogue without mediators and referees. We hope that we can nudge the dialogue in the direction of a cooperative courtship.

Since so much of this book is about various evolutionary paradigms, we begin with an extreme example of dichotomy, which is far from our ultimate goal of reconciliation. Although we are loath to highlight such overt conflict, the case in question has an interesting historical context. We refer to Richard Dawkin's provocative book *The God Delusion* in which he launches preemptive attacks on religion so strident that Michael Ruse, our normally unbiased and admirably philosophical dialogue mediator, was moved to declare, first that Dawkins made him ashamed to be an atheist, and then to state later that the book was one of the worst he had ever read. [47] Meanwhile, Alistair McGrath who at one time said he had avoided challenging Dawkins for more than twenty-five years finally responded with his critical reply *The Dawkins Delusion*.[48] What makes the Dawkins-McGrath debate extraordinary is not that it is a 'low level' confrontation between a fundamentalist creationist and a zealous scientist, but instead it is a pitched battle between two Oxford academics with impressive credentials. Curiously this happened once before, in June 1860, when Oxford staged the historic Evolution versus Creation debate between Thomas Henry Huxley and Bishop Wilberforce. Because Darwin recoiled from public debate, and was well aware of how his theory might offend religious sensibilities, Huxley spoke for him and gained a reputation for being Darwin's 'bulldog' and for disingenuous, politically-motivated attacks on religion.[49] History will probably decide that the current Oxford squabbles are just as unnecessary as those of the 1860s.

Many have wanted to reconcile science and religion, not least because there are many scientists, like Alister Hardy, another Oxford biologist, who overtly declare their religious convictions. Clearly such conciliatory objectives are well motivated if not admirable. But, in our opinion, it is necessary to understand what advocates mean by science, religion and spirituality in the first place. We have already stated the position that so-called 'post-religious spirituality' seems to resonate better with the inclusive nature of the integral consciousness paradigm than with narrower exclusive religious positions. But the same spectrum of exclusive-inclusive worldview also occurs in the scientific community. Thus, there is a great difference between the cosmologist, so awe-struck by the grandeur of the universe that he includes God in his book titles, and those like Dawkins who defend entirely different, and rather narrow definitions of Science. In a nutshell, the whole science versus spirituality debate is, by definition, severely limited if either science or spirituality is too narrowly conceived. It is as if a referee were forbidding us to bring the full range of our consciousness to bear on the problem.

Perhaps a key issue, too, is that a narrowly defined science versus traditional, religious (theistic) dogma problem is old hat and perhaps intractable, whereas the integration of broadly defined 'philosophical' science and a more 'progressive' spirituality is an entirely different and potentially more interesting dialogue. Perhaps quite simply it is time for the rules to change, not by arbitrary decree, but by the intrinsic dynamics that have always generated paradigm shifts: that is, global mind changes.

The spiral of thesis, antithesis and synthesis allows us to rescue the baby from the bath water and dethrone old and jaded paradigms with efficient new consciousness structures that allow for the reintroduction of awe, wonder and the sacred into science.[50] As such new consciousness structures emerge, so too will newly-minted paradigms, quite possibly with surprising characteristics. Michael Polanyi has stressed the importance of the exceptional intuitive powers of original scientists who challenge and dethrone the orthodoxy of existing paradigms.[51]

If science makes the mistake of being magisterial and defining itself too rigidly, in terms of quantitative reductionist methods and rejection of opposing heterodox views, it will paint itself into a corner. It will do what religion (theology) did as modern rational science rose to prominence during the so-called Enlightenment.

There will always be controversial borderland domains of science beyond the reach of immediate comprehension and simplistic explanation. It seems that knowing ourselves is no easy matter and almost certainly a quest that has no absolute end. At present, consciousness studies, parapsychology, and other subtle energy domains of psyche and spirit tend to occupy the borderlands of comprehension, but this is precisely the juncture that the expansion of science and human inquisitiveness has reached at the present time. Whether the ability of our evolutionary consciousness to inexorably penetrate these domains will lead, with hindsight, to claims of a new chapter in the history of science, rather than an entirely novel, newly-labelled structure of integral consciousness remains to be seen. Perhaps, our bias is that the vocabulary of the integral philosophers already has a foothold. But there is no guarantee that this is or will be the case.

Goethe held the perhaps poetic, but nonetheless sensible view that one could not approach nature with the objective of defining and constraining her with rigid 'natural' laws without to some extent falling under the spell of these incomplete abstractions (laws), and overlooking and denying the mysteries beyond these laws. Barfield restated this subtle but nonetheless very real problem of self-delusion with the metaphor of idolatry.[52] Believe too much in the world, the symbols, icons and abstractions you use to represent the world to yourself and you have already fallen under their spell. Similarly, Goethe warned that one should not conclude ones scientific investigations with the misconception that fixed laws have been 'determined' unless one is willing to fall under their deterministic spell. Some of this may, on reflection be self-evident, as history tells us that no paradigm has proved inviolate, and even the most robust theories will inevitably require modification. Paradoxically, therefore, history demonstrates that theories and doctrines can be both stepping stones and obstacles. Those who are ahead of their time may propose innovative theories that are ignored for generations or centuries before being rediscovered and affirmed. Goethe, for example, challenged Newton's theory of light by saying that it was incomplete, and to this day there are still those that argue in favour of Goethe's view.[53]

Ultimately the constraints on our science or spirituality are those imposed by the limitations of consciousness at any given time. Any law of scientific limitation can potentially be dethroned. Psycho-spiritual consciousness has the potential to transcend itself continuously. As Polanyi notes, a society refusing to be dedicated to transcendental ideals (truth, conscience and freedom) chooses to be subject to servitude, ideology and the dictatorial power of institutionalization.

Longing for love: the happiness of homecoming

We hope it is clear that in attempting to integrate eastern yogic concepts with certain paradigms of western science, we have no intention of advocating that esoteric practices associated with so-called mysticism, transcendentalism or any other new paradigm 'isms' replace good old empiricism. On the contrary we only suggest that it is necessary and perhaps inevitable that the 'schisms' and 'isms' come together and learn from one another co-operatively. There is good evidence that evolution works by cooperation as well as by competition.[54] All we have learned about the homeostatic balance of nature tells us that cooperation is fundamentally important to the continued health and survival of ecosystems, even though, paradoxically such systems need antagonistic tension.

Cooperation involves communication or communion between individuals, species and cultures, and in the evolutionary literature, both biological and psycho-social, we find such communion also referred to as altruism and even love. Perhaps surprisingly, in his book *The Ascent of Man* Darwin used the word Love almost one hundred times and stated clearly that, in comparison with animals, humans lived by rules of cooperation and altruism quite different from those governing natural selection (survival of the fittest). David Loye dubbed this *Darwin's Lost Theory of Love*.[55] It is therefore rather ironic that sociobiologists, like selfish gene and meme advocates, incorporated the very human notions of altruism and selfishness into the natural selection paradigm, without considering how much they were projecting human psychological bias into their models. .

So what does it mean that Darwin's ruminations turned from biology to love when he considered human nature? Why is it that love becomes the ultimate theme in the writings and pronouncements of all the great mystics? Why are true spiritual leaders like Buddha or Christ, and even, arguably, more secular leaders like Gandhi and Martin Luther King, revered for their compassion and non-aggressive philosophies? Do they know that ultimately, despite war and petty divisions, the meek will inherit the earth? Why does our vocabulary speak in superlatives of higher consciousness, the higher Self and Superego? Ostensibly the material rewards for the selfless, mystical and spiritual way are often negligible. Indeed, they may all too often be exile, persecution and death. So, as the cynics say, the 'good guys finish last'. In an enlightened society, preaching love and compassion is to be welcomed, but all too often powerful forces are marshalled against the authentic spiritual leader. We see this theme reiterated again and again in our literary epics. The peaceful life of simple good people is disrupted by dark, irrational and dangerous subconscious undercurrents, and from nowhere a Hitler or Darth Vader bursts on the scene in a land previously renowned for its poets and philosophers. In such cases chaos overtakes order, and what began with mysticism ends in politics. But thankfully, some epics have happy endings and a hero, in the guise or a wizard or hobbit, often with a little mystically inspired supernatural guidance, reverses the dark tide and we all cheer as the spiritually superior forces win the day, finally allowing mystic faith to trump politics. Why, in biological and evolutionary terms, do such epics resonate so deeply in the human psyche?

But why in the first place do these dark forces array themselves with such vehemence against peace and compassion? The reason is that a shift in consciousness has remarkable transformative power. Whether the reactionary response is conscious or not, it must be powerful to confront the transformative agencies. It is perhaps for this reason that commentators like Karen Armstrong and William Irwin Thompson suggest that axial ages of transformation and renaissance are accompanied by dark ages.[56] As the light intensifies, so only the darker places provide a refuge for those resistant to transformation.

It really does seem that humanity's history is one of spiritual struggle with our own disorderly psyches. Here the 'chaos and order' metaphor seems appropriate. Just as the emergence of reason (mental consciousness structure) brought a measure of law and order to a world of emotional turbulence, so cosmic consciousness brings the promise of hope and a more compassionate new age post-religious spirituality to a world of intellectual and ideological confusion. As the sages remind us, the transformation can be painful and involve protracted painstaking work in art, science and spiritual questing. In order to gain something, something else must be lost: Superego dethrones Ego. One must break eggs to make an omelette, and crack the shell to reach the kernel. In systems jargon, transformation operates through the interplay of anabolic (constructive) and katabolic (destructive) forces.

For all the thousands of anthropological and sociological books that examine the dynamics of culture, there are relatively few that treat 'love', charity or compassion in any detail or from an evolutionary perspective. As noted in Chapter 7, a well known evolutionary theory of symbiogenesis holds that cells with a nucleus, known as 'eukaryotes', evolved from 'prokaryotes' (cells without nuclei), as a result of a merging or union. Perhaps this was the biological origin of love? Certainly in terms of biological definition it was the origin of sex, because asexually reproducing forms evolved the capacity for sexual reproduction. This shift from mitosis to meiosis meant that the asexual ancestors were no longer immortal clones. Each parental pair in the new sexual generation gave part of its genetic make-up to the offspring to produce something new: a 'synergetic' offspring that was greater than the sum of the two parental parts. In doing so the older generation was genetically out of date, and no longer immortal. This great biological phase, shift which took place some two or three billion years ago, has been described as the origin of sex, but also as the origin of love and death.[57]

Curt Thesing's *Genealogy of Love* echoes this bittersweet love story.[58] In looking at the sex or love life of a range of organisms from amoebas to mammals, leaving the final chapter for humans, he notes that: 'the folly of love, this mingling of pleasure and pain, restlessness and joy ... periodically affects ... the whole world of living organisms' — especially in the prime of life and in the springtime. Here we are reminded that 'love' is a powerful elemental force associated with the cyclic rebirth of the entire biosphere. Perhaps this would explain why the timing of the psychological rebirth which Bucke labelled as cosmic consciousness also appears, according to him, to be primarily a springtime phenomenon. Thesing stresses the absence of information that would shed light on the love-life of our earliest ancestors. Cave paintings show the physical act of lovemaking, which he says fulfils 'nature's supreme purpose' of procreation, but this tells us nothing of the psychological aspects. So Thesing turns to the classic, if archaically titled, work of Malinowski on *The Sexual Life of Savages*. Among the matriarchal Trobriand Islanders, 'blood relationship is traced exclusively through the mother ... they do not recognize that the father has any part in producing a child.'[59] This is exactly the conclusion reached independently in Chapter 5 above, to account for the matriarchal, mother goddess, organization of ancient archaic and magic consciousness cultures in humanity's infancy (Keck's Epoch I).

Instead of starting with amoebas, Sydney Mellen takes up the subject of the *Evolution of Love* with hominid history, attempting to uncover love's biological origins and explain this 'powerful force', the 'least well understood of our capacities', in a Darwinian or sociobiological context.[60] He suggests durable emotional attachments between men and women may have marked the 'beginnings of love' as early as two million years ago, when supernatural (that is, proto-religious) beliefs were 'imagined' to account for an external world that was difficult to understand. This rather outmoded view is seriously at variance with the explanations of Long and Barfield (Chapter 8) and says little of the rich historical and literary record so cogently interpreted by Morton Hunt in *The Natural History of Love*.[61] Dealing only, but most thoroughly, with western history, Hunt considers that primitive love hardly resembles western love at all. He tells of a primitive tribe that could not understand a European fairly tale involving an intrepid lover who overcame near-insurmountable trials and tribulations to obtain the hand of a fair maid. The chief simply asked, 'Why not take another girl?' Personal relationships are not valued where individuality is little developed.

Hunt places the origin or 'invention' of love with the Greeks, who had two concepts: *eros* and *agape*, for carnal and spiritual love respectively. Today we still make the similar distinctions between lust and love. By modern standards the Greeks of the Axial Age, around 500 BC, were debauched, male chauvinists. Lusty bisexuality, homosexuality, prostitution and abortion were rife. In the new urban settings, family was less cohesive than it had been in tribal Homeric times (1300–1100 BC). Marriage had little to do with love, being more a civic duty for procreation. Where wives and husbands were less valued than blood relatives, the state worried about population decline. Public esteem and the avoidance of shame far outweighed any sense of internal guilt or ethics.

Among the Romans, lust was also rife and undiluted by any sense of sin or the need for fidelity. Julius Caesar was a perfumed, bisexual lecher who indiscriminately loved the wives of his friends and most of his political associates: he was known as 'the husband of every woman and the wife of every man'. Love was seen as an emotional affliction and deluded 'madness in which the lover longs unendurably to have a complete union with another body although complete union is forever impossible'. Love was just an urge to be satisfied, an emotional feeling to be conquered by any means available, including, argument, rejection or tricking of partners. In such psychological circumstances it is surprising that the value of women was actually increased by the cultural rules governing dowries. A woman could leave a difficult husband and take her dowry with her, unlike her Greek counterpart she was free to leave the house and engage in social intercourse, in a society where at times adultery was quite fashionable and one made liaisons not with 'a' person, but with 'any' person![62]

In what Hunt calls 'the dark ages of love', ... 'the Roman Christians had dissociated love and sex, the former being God's business, the latter being the Devil's'.[63] The emotional tone was changing, what had been a guiltless habit, became a guilty one, and although one could pay a penance for indiscretions, marriage was the least guilty way to enjoy sex and obey God's command to procreate. Even though this helped rebuild the family, sex was still considered shameful and many fled to the celibate life of monasteries and convents. In a single terrible sentence, Saint Augustine epitomized the profound Christian conflict: 'Through a woman [Eve] we were sent to destruction; through a woman [Mary] salvation was sent to us'. But amid all this ambivalence the new and powerful notion of romantic love arose.[64] Perhaps we might again cheer as it seems that eventually love, or at least a new species of love 'conquers all'.

As the Dark Ages gave way to the medieval age of renewal, a sea change was underway. As mentioned in Chapter 8, so-called 'courtly love' or chivalry introduced a form of etiquette in the service of women that has shaped western manners and morals to the present day. Although the proverbial knight, who may well have been married, often went to ridiculous lengths to earn the right to adore a chosen damsel from afar, his adoration of an ideal woman represented an uplifting spiritual love with the novel 'monogamous' component of a one-on-one devotion that suppressed his lust, and showed him a true Christian. How different, one wonders, was such adoration of an idealized princess, from the mystic's adoration of Mary or Christ? Changes between medieval and early Renaissance art show the image of woman, and particularly the Madonna, transforming from saintly, two-dimensional, gilded icons into sensuous three-dimensional women with exposed flesh and vital, individualized personalities. Yet paradoxically, while Renaissance artists glorified women, at this same time they suffered far greater persecution as witches than they had during the Dark and Medieval ages. Sadly, the price for the emergence of the sensuous woman in some quarters was brutal persecution in others. Without our pointing fingers, the same dynamic plays out today in certain cultures where religious oppression is still rife.

As we march up through the Enlightenment and into Victorian and modern times, the one-on-one romantic relationship became an iconic, pivotal part of literature and a vehicle for examining individuality and character. Puritanism generally endorsed sex and procreation within marriage but was strict and heavy-handed in its punishment of adultery. Victorian manuals, written mainly for women, typically give advice on creating peaceful domesticity and raising 'the tone of her husband's mind and leading his thoughts to dwell on a higher state of existence'.[65] Thus a good woman could help spiritualize a man, even in a society where repression had no small influence in giving rise to a significant increase in the prostitution.

Hunt concludes by characterizing the Modern Era, up to 1959, as 'The Age of Love' and stating that 'at no time in history has so large a proportion of humanity rated love so highly'.[66] Freud, of course, taught us that the unconscious was filled with sexual urges, but even without knowing his work and the insinuation that sexual longing resided in the deepest recesses of our being, the average citizen could not escape movies and magazines overflowing with love stories and pictures of sexy bodies paraded for mass consumption. Intellectuals advocated open marriages, later called 'free love'. Feminists championed the female orgasm and the right to chose, and chivalrous encounters morphed into dating and sex outside the institution of marriage. Had Hunt completed his excellent study ten years later, he would surely have had to include the liberated Sixties in his 'Age of Love'. In this turbulent, but ultimately optimistic decade, the message of the so-called 'love generation' intensified, beginning with the romantic 'Love, love me do' and spreading globally and philosophically into 'Love is all you need'. Moreover, in the spirit of communion, mass musical and multimedia rallies were held to celebrate 'love-ins', togetherness (and sex), and most germane to our theme, the love generation also named itself the consciousness generation and espoused allegiance to peace, mysticism and spiritual ideals. For the first time in history, large anti-war rallies commanded global attention.

Here it is worth adding that human-animal relationships have undergone profound transformations during the two millennia since the days of the Roman games. Life was then cheap and along with human victims, hundreds of thousands of animals were regularly slaughtered in giant bloodbaths. On some deep level, the Romans felt that inside the Coliseum they had control over the savage and chaotic forces of nature that ran rampant outside. Like love, nature was a force to be conquered. Even in Victorian times our most distinguished naturalists, like Darwin and Wallace, thought nothing of killing multiple specimens of rare species for museums, and hunters drove the ubiquitous passenger pigeon and many other species to extinction, and nearly did the same for the buffalo. Although horses, dogs and other domestic animals had 'value', even in comparatively modern times, it was a relatively rare voice — as in William Blake's lament: 'A robin redbreast in a cage, puts all heaven in a rage' — that noted the abuse of most wild creatures.

Even in Darwin's day, educated Victorians expressed disgust at the thought that we might be related to savage jungle apes. At a time when the Bushmen and other tribal Africans were still hunted like animals and killed for sport or to be sold into slavery, no one shed a tear over the loss of a gorilla or chimpanzee for body parts made into trophies. Compare this situation with the present day, when the same bourgeois western intellectuals will give money to help conserve the apes, studied by leading scientists who love them, name them, provide them nature reserves and immortalize their family histories in best-selling biographies. Increasingly we abhor the use of any animal for laboratory vivisection, and we routinely rescue mangy stray cats and dogs from animal shelters so that our loving children can adopt them as pets. Our love of our fellow creatures has come a long way since Roman times.

In *The English Spirit*, D.E. Faulkner-Jones' profound excavation of our literary heritage, we are reminded of Rudolf Steiner's great insight into what he called the Sentient Soul, the Intellectual Soul and the Spiritual Soul, corresponding to his third, fourth and fifth 'post-Atlantean' epochs. He saw these faculties typically emerging in individual development (ontogeny) around 21, 28 and 35 years respectively, and Faulkner-Jones gives wonderfully perceptive examples of their literary manifestation in the careers of many of our greatest writers from Shakespeare and Milton to Wordsworth and Tennyson. Beyond the obvious resonance with inherent seven year cycles (see Chapter 3), on a collective cultural level there are striking parallels with Gebser's mythic, mental and integral consciousness structures (coincidentally the third, fourth and fifth in his schema, as outlined in Chapter 5 above). This evolution of soul also parallels Barfield's three stages of participation, separation and final participation. The Sentient Soul participates in a deep feeling of communion with nature, but its shadow side is weakness of Ego and individuality. The 'thought-dominated' Intellectual Soul stands outside nature seeing 'a universal divine principle permeating the material universe', which some call natural law.[67] Such intellectual abstraction may generate a disquieting sense of loss and the shadow of morbid introspection, as well as the hollow belief that knowledge can only be gained by sensory experience. The Spiritual Soul, however, slowly becomes inwardly filled and more reverentially conscious of the external world, adopting a new sense of proportion characterized by devotion and temperance. In short, it is capable of true love, manifest in a compassionate understanding of others as divine spirits constrained by the limits of physical incarnation and mortality. To love is to see the divine individuality in others.

According to Steiner and Faulkner-Jones, the emerging love forces of the Spiritual Soul epoch are just beginning to manifest in human culture. In psychological terms, collective intellectual thought-forces slowly descend into the individual Ego to be integrated, owned and balanced with feeling and will. This view is remarkably and intelligently convergent with Gebser's insights and other notable spiritual teachings. Faulkner regards such profound insight into human psychological evolution as an entirely rational take on the evolution of love and consciousness, widely supported, if data is needed, by our very best literature.

Perhaps it is a conceit of mental consciousness optimists to suppose that we live in the best of times. Our material standard of living is unsurpassed and we are generous enough to wish that others share this luxury, even if we sometimes wonder whether we do enough to actively help our neighbours. We genuinely believe that equality, brotherhood, love, liberty and the pursuit of happiness are legal and moral rights that should be fostered and maintained as part of the social order, and perhaps this will force their intensification in the future. As a result, we pursue these goals in many ways: selfishly for ourselves; somewhat more altruistically for friends, family and community; and sometimes with genuine unconditional charity for the benefit of suffering humanity. What then does the history of love and our present understanding of this complex phenomenon, tell us about our spiritual evolution and the relationship between Id, Ego and Superego?

Clearly, the benefit of historical hindsight suggests that love was once entirely a latent force subservient to the lusty, Id- and loin-driven animalistic sex drive. There is, no doubt that these erotic (Eros) drives are still prevalent in many quarters, but as emergent consciousness structures have influenced (awakened) the higher organs (heart and brain) so too the Ego has emerged with novel consequences for the individual and the institution of love. As developing emotion and intellect added new dimensions to individual personality and character, richer communication was possible and individuals could entertain romantic feelings of affection for others that involved chivalrous and respectful sensibilities that appeared to reach beyond the physical body to the intangible, inner, soul and spiritual qualities of the other. Since the time of Plato and Plotinus, idealized notions of a higher spiritual order characterized by beauty and goodness, have progressively exerted a powerful influence on our emotional, intellectual and spiritual sensibilities. And for much of this time, in the West, once Christianity gained its foothold, God and the 'good' became more or less synonymous. But as earlier sections of this chapter indicate, the notion of higher orders of spiritual consciousness had been independently and elaborately developed in the East through a type of physiological practice, and may well have had more of an influence on western spirituality than we know. The essential point is that evolution appears to be all tied up with the emergence of higher or more intensified states of consciousness that repeatedly stress the importance of loving communion between self and other, whether that other be a child, parent, friend, lover or some manifestation of the natural or supernatural order.

Although the eros-agape distinction between carnal and spiritual love has long been recognized, there is no evidence of a trend towards complete or permanent separation, although repressive religious cultures still exert strict institutional control over sexual behaviour, as they did in the past. Most modern sensibilities regard the blending or reunification of sexual and spiritual love as a path to health and happiness. The orgasm is to be enjoyed as both a physical and spiritual communion between loving partners. The ecstatic nature of full organism transports the lover to another state of consciousness where he or she feels the dissolution of Ego and fusion with the other. The French call it *la petite mort*. Likewise, this little death of the Ego is experienced by the mystic when they feel the ecstasy of unification with divine spirit. Although having nothing to do with carnal sex, the mystical union is all about loving communion between the individual and God (Godhead), or as it is sometimes termed, a marriage between bride and bridegroom. Likewise, in attaining states of higher universal consciousness, variously named by adept mystics and yogis, the individual reports sustained states of bliss and a knowledge, or direct experience of Christ Consciousness, the Buddha nature, *moksha* (liberation), the fusion of Atman (the human soul) with Brahman (the indescribable, Supreme Being) and so on. Most of those describing intense spiritual experience report a feeling of great happiness and wellbeing, as if bathed in love or touched by grace. As we know, Bucke described this as a recurrent and significant tendency towards enhancement of the intellectual and moral faculties. This sense of communion with the higher Self or Superego is also reported by those experiencing the rising of the kundalini. Indeed, Krishna simply calls this the evolutionary energy in man, suggesting, as others have done, that: 'This is the next stage in the evolution of our nervous system, and it is a necessary correlate of spiritual development ... toward which all mankind is moving'.[68] This potential exists in all humans and is seen as a latent force that will awaken progressively with time.[69]

Hopefully we have said enough in previous chapters to convince the reader that the activation of new physical and spiritual organs, is part of the ever-changing, biologically or physiologically based dynamic of consciousness. We also hope that the argument that, as a whole, humanity is moving towards an intensified, and more universal, cosmic or integral structure of Superego, is supported by the historical evidence. This trajectory involves intensification of communication and communion including what we have labelled as love. The emergence of spiritual love from carnal love, followed by the reintegration of the two seems to mirror Barfield's original participation, separation, final participation model, and could also be a good metaphor for much authentic life experience.

Mystics and non-mystics alike long for the happiness, ecstasy, love or the sense of grace that comes from the authentic spiritual experience or happy communion with other sentient beings. The reason is that we long for spiritual reunification with the universe from which we separated. This has little to do with unsubstantiated religious belief. We now know enough about human ontogeny and phylogeny to assert with some confidence that humans only came into being, as conscious psychological entities, after a long period of dormancy during which time we were part of the unconscious psychic web of the biosphere, and indeed of the cosmos. The conscious tributary has only recently entered the unconscious river, but in doing so the current has changed in quality. The awakening of the conscious soul in the individual is but a replay that parallels and quickens the awakening of all humanity to self-consciousness. Only in this state can the individual or species ask 'Quo vadis?' — hoping that directions and destinations may exist. Likewise, throughout history, consciousness gurus have indicated that answers are found when consciously crossing the 'next' threshold to cosmic or universal consciousness. In doing so, the sense of separation evaporates and the experience is one of coming home, being at one with the cosmos and being immortal. Deathbed and near-death experiences further suggest the reality of a spiritual afterlife, and moreover are very often associated with deeply moving feelings of unconditional love.[70] If one is not too cynical, then it is also perfectly reasonable to interpret the many religious and spiritual beliefs in immortality as manifestations, not of wishful thinking or mental delusion, but of authentic efforts to interpret the collective psychological intimations of humanity regarding our true origins. For simple consistency we must remember that the ancients, like the modern shaman, report the experience of emerging from the universal collective unconscious, while yogis and mystics report transcending the dual world of sensory experience to consciously 're-enter' universal consciousness. So, much evidence allows us to infer internal consistency in a growing number of empirical reports, and conclude that as individuals and species we are inseparable from the unfolding of an immortal biosphere. Our experience of life and death, like waking and sleeping are only manifestations of evolutionary changes in form and consciousness.

As noted in Chapter 8, the Jesuit paleontologist Teilhard de Chardin predicted a future species — Omega man — whose spiritual consciousness was evolved far beyond ours. He believed that the development of human personality made possible the highest human activity, namely communion or love. In *Homo sapiens*, says Teilhard, spirit evolved far enough to reach self-consciousness and differentiate from the biosphere. But after this differentiation (separation), lines of spirit again converge, as seen in the process of human socialization and the globalization of science. This process requires that humankind go beyond knowing itself as an individual member of planetary society and 'acquire the consciousness, without losing themselves, of becoming one and the same *person*'.[71] If this is construed as a global brain type of communication, then it parallels the type of conscious psychic communication suggested by Barry Long as the next or final participation. Teilhard suggests that the only word for this type of convergence is love — 'the most universal, mysterious and tremendous of forces — the primal and universal psychic energy, as modern psychology likes to stress ... Love is a sacred reserve of energy — the blood of spiritual evolution'.[72] Teilhard believed that convergence of all forces in the noosphere would ultimately lead to a 'dynamic of unity within science, religion and society'.[73] With this convergence on Teilhard's Omega point, we reach a union of spirit with 'God' marking the apparition of the 'theosphere'.[74] Using strictly paleontological language, he even described love as a new phylum. Certain of Wilber's hierarchies endorse this scheme with similarly nested Physiosphere > Biosphere > Noosphere > Theosphere schemes.[75] (See Figure 9.6.) Although Teilhard was a Christian as well as a scientist, his mystical insight and arguments for a goal-directed straight-line evolution or 'orthogenesis' have fascinated, frustrated and intrigued philosophers, scientists and theologians alike.[76] Even physicists like Frank Tipler have got in on the act, formalizing Teilhard's ideas in *Omega Point Theory* which purports to provide mathematical proof of just such future convergence.[77] The acronym OPT is delightfully apt, for it derives from the Latin *optare* meaning to choose, particularly with regard to citizenship and free will. But as this is not a mathematical treatise we will leave Tipler's extraordinary ideas for another day, and note only that Tipler's theory label pays an unusual, mathematically-oriented tribute to Teilhard's vision.

We need only add Teilhard's name to the list of persons who predict the next stage in evolution as a 'post-religious spirituality' in which love and communion become important unifying forces. He wrote:

> Some day, after mastering the wind, the waves, the tides
> and gravity, we shall harness the energies of love and then
> for the second time in the history of the world we will have
> discovered fire.[78]

Likewise Johannes Rohen concludes his groundbreaking book on the dynamic wholeness of the human organism with a strong endorsement of Orthogenesis: that is, purposeful evolution towards a goal. He notes that: 'The fact that the human brain still has many areas with no

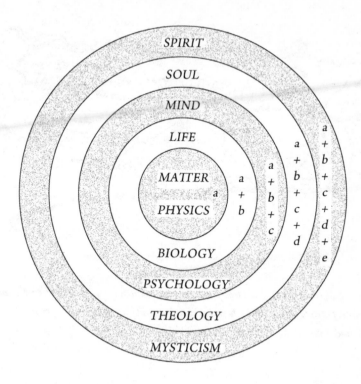

Figure 9.6. Wilber's hierarchy of the physical, biological, psychological and theological /spiritual worlds is similar to Teilhard de Chardin's levels of physiosphere, biosphere, noosphere and theosphere.

assigned functions is an important indication that the human race will continue to evolve to higher levels in the future'.[79] (See Figure 9.7.) We are partly empty vessels waiting to be filled. This sentiment echoes Jos Verhulst (see Chapter 7). Like others in the anthroposophical tradition, Rohen understands that the oft-cited reference to humankind's 'fall' is not an abstract religious metaphor but an 'empirical' reference to the process of self-consciously coming into being. Likewise Rohen unashamedly uses 'love' as a term to describe creative forces that shape evolution, and he also reiterates Teilhard's now famous quote cited above. Given that the glimpses of higher consciousness given us by

Figure 9.7. A remarkable Inca fresco showing an evolutionary 'chain of being' from worm and pre-wormlike beings to apparently angelic humans. (After Verhulst 2003, also Heuvelmans and Porchev 1974; from Vollmer 1828.)

mystics and spiritual gurus consistently speak of the power of love, it is a clearly a force to be reckoned with in any scientific or philosophical investigation of consciousness. As many a recent scientific trend indicates, our own dynamic consciousness is leading us into the 'post-physical' or 'post-material' realm of subtle energy disciplines: among them information, communication, coherence and now love.

When writing about human and cosmic evolution, many a scientist has used the metaphor of 'our place in nature' or 'our place in the cosmos'. This connotes more than the simplified notion of a physical and spatial location, on planet Earth, in the Milky Way galaxy. The fascination with our 'place' in the grand scheme of things has much broader psychological and spiritual implications. We are of course at home in the cosmos, and the psychological urge to find our place is perhaps no more than an indication that during our fall into self-consciousness we lost our psychological bearings in the excitement of discovering the external sense-perceptible cosmos. The same forgetfulness came over us after we were born and got so caught up with the adventure of sensory existence that we had to ask other adventurers where we came from, and soon found they too had mostly forgotten. But the search for the way back is, on the one hand, an illusion, because we are already there: that is, here! 'No matter where we go there we are.' On the other hand, the whole dynamic cosmic and organic flux is, as near as we can tell, not only simultaneously here, there and everywhere; it is also the process that creates us physically and psychologically. We have floated on the primeval oceans of the unconscious and subconscious with fellow sleepers and dreamers to find ourselves awakening at home on the wave of self- consciousness. But the currents are subtle and powerful, and the journey has just begun.

Postscript

This book is the product of *only* two like-minded but nevertheless quite distinct human consciousnesses among a sample of several billion, but we have benefited from many influences. While each individual consciousness is subtly if not substantially different, we also each belong to communities that share common, cultural and psychological traits. Moreover, as our sage guides have suggested consciousness is ever-changing and the source of novelty is ever-present. Thus, the child's self-consciousness emerges as suddenly as does the unexpected mid life epiphany, the emergence of high Paleolithic art, a new consciousness studies paradigm or perhaps even the mysterious appearance of a new species. It is in the nature of novelty that it may erupt from its ever-present wellspring at any time

What many have characterized as time's arrow and the progressive trend towards larger brains and perhaps a more evolved and more ethical consciousness is not necessarily achieved by a steady, gradual march towards a better, utopian world devoid of setbacks, pitfalls or psychological and spiritual challenges. Our final chapter suggests love, cooperation and all that 'good stuff' as the progressive evolutionary way, but that is not to say that authentic love is as easy to realize as its many alternatives. G.K. Chesterton thought the idea of a 'slow movement towards morality' and ethical progress, potentially naive and liable to encourage laziness. It perhaps does not so easily accommodate the unexpected lunatic like Chesterton's St Francis who unpredictably 'jumps up' to protest an injustice, or soar aloft on a transcendental electric current to declare he has been born again in the realm of a new consciousness structure. Nor does it easily accommodate unexpected dark forces marshalling ominous war clouds beyond the familiar horizon. Emanuel Swedenborg (1688–1772), who underwent one of the best documented consciousness transformations in modern history, suggested that divine providence allowed everyone to act in freedom, so that 'unless evils were allowed to break out, we would not see them, and therefore would not admit to them, and so we could not be moved to resist them.' He did not foresee an end to all conflict. Nevertheless, despite the play of evil, Swedenborg believed in a better future.

Activists rightly remind us that 'excessive confidence' and certainty smack too much of a modernist, colonial hubris, or what academics call 'foundationalism'. None of us should be certain we are always on the right course, lest we stick our head in the sands of complacency. To profess certainty, or idolize what Barfield called the 'residue of unresolved positivism', is akin to arguing that only our way is right, that all of the people in 'our team' can in fact 'be all right all of the time'!

We are not so credulous or unaware of the dark side that we subscribe to the mental consciousness doctrine of steady, linear progress and the too complacent hope that all will turn out 'just right' in the end. We would soon be bored in such a world. Our aim has been to discuss the dynamics of consciousness and not offer a comprehensive philosophy of existence. Undoubtedly the dynamics of ontogeny, history and prehistory (phylogeny) indicate something of a roller-coaster ride. The steady and deliberate climb to a higher vantage point represents, if you like, a constructive and rational *faber* phase of progress. Then suddenly, the journey becomes a terrifying emotional plunge into chaos. The rhythmic up and down, the spiralling dance of the whirling dervish, twirls us in and out of historical order and hysterical chaos that all our language and scientific philosophy can only partially explain. We are all along for the ride and perhaps our highest challenge is to be steadfast travelling companions.

One can serve the cause by writing optimistic books about purposeful journeys towards enlightenment. Equally if not more valid than academic discourse is the 'action' of those who 'jump up' with postmodern rallying cries to banish complacency and help the disadvantaged and oppressed, with authentic engagement in their lives and causes. Such aspirations seek meaningful communion, right direction, harmony with a higher order and relationship with the *mysterium tremendum,* whether seen, glimpsed, intuited or unseen. But perhaps as Swedenborg suggested, we should always be conscious helmsmen, constantly seeking the right course, which in turn is only discovered by seeing where 'evil breaks out' and responding appropriately.

Our best science and theology evidently agrees that human self-consciousness is a latecomer on the 3.5 billion-year-old evolutionary stage. Likewise all our best literature, psychology and spiritual traditions, suggest that this awakening involves investigation of our own inner demons and angels. The gift of self-awareness allows us to personally excavate the frontiers of the known to explore the murky realms of the unconscious. But as we dig, so the collective consciousness of humankind is also illuminated, promising to better inform our identity and shed light on our proverbial place in the universe. To say that this archaeological project involves 'nothing more' than bringing humanity's unconscious into consciousness, is surely to underestimate the enormity of the task. But once a species becomes self-aware, and this is indeed our collective psychological reality, charting our destiny and understanding our deep time deep-time origins is indeed an epic evolutionary undertaking. In one spiritual tradition, the Buddha dug down through a billion generations of organic evolution to experience the essence of the first cell, the first gene, and the stardust from which these originated. Surely our psychological evolution as a sentient species destines us to follow in the illuminated footsteps of Buddha, Christ and other spiritual guides as we excavate the vital dust, the very fabric of Mother Earth? And beyond, as mind penetrates matter, is not the dark, unconscious void of interstellar space, where we project our aspirations, already illuminated, sentient and pregnant with the possibility of existence?

Our elusive consciousness is destined for the psychological adventure of existence and self-awareness in each individual incarnation. But this destiny, the challenge of explaining the world (the other) to ourselves, necessarily involves relationship. As the eye sees other eyes, and the I sees other 'I's', so we create our individual and collective realities and sacred spaces. The familiar metaphor of 'exploration' applies to the embarkation point of self-consciousness sitting like a shining peak atop the iceberg of a sentient biosphere. Submerged in the dark waters below, ancestral organic relationships have a subconscious and unconscious pedigree that reaches back to life's origins. Hearts and minds feel and resonate with other hearts and minds. The eye and the I may each experience itself in myriad forms as a physical body, a biochemical flux, a rational brain, a dreamtime archetype, or even as a flickering field of electromagnetic light-speed energy integrated with the cosmic continuum. As in Anthony Freeman's (2003) definition of consciousness, we are mysterious structures in space that somehow hold ('encode') knowledge, mere will-o-the-wisp 'forms', building and rebuilding like gossamer in a creative dance of subtle energy infused with elusive flights of consciousness.

Such 'knowledge,' or should we say 'creative sentience' may manifest as our experience of happiness, sadness, loneliness, enthusiasm or ecstasy. We may experience others, or indeed ourselves, as angels, demons, lovers, mentors, saviours or frustrating challenges to personal growth. Some human brethren are literally brothers and sisters in flesh, others are soulmates or spirit guides, but rarely are any two experiences the same.

At the heart of the matter is surely a quintessential instinct that we are part of a deeply intricate and dynamic experiential process, beyond the full comprehension of even our most refined senses. We may grope for words like 'existence,' 'sentience', 'consciousness,' 'being' or 'becoming,' and we may have flashes of illumination and insight that cast a little light on the ever-elusive mysteries of our sentient, intellectual and spiritual souls. But right before our eyes is the perennial manifestation of a divine creativity, of an awe-inspiring intelligence. What is more, we are deeply embedded in the universal mystery and tap into the vital energies which at every instant provide the very life blood of our nourishment, inspiration and being. We enjoy the gift of communion with this universal consciousness and we are equipped to pay homage with our subtlest poetry, the most elegant mathematical theorems, our most sublime musical creations and our acts of love and kindness. What could be more heartfelt, adventurous, innocent and purposeful, than to surrender our creative human spirits to the cosmic *mysterium tremendum*? Like wide-eyed children, clutching our tickets to ride, our ever dynamic, organic, pulsating, cosmically conscious world awaits our full participation with a promise of hundredfold rewards and as yet untold adventures.

Endnotes

Introduction

1. Two recognized sources on the phenomenon of the release of kundalini energy are: Krishna, Gopi (1997) *Kundalini: the evolutionary energy in man,* and Sannella, L. (1987) *The Kundalini Experience.*

2. Spinoza's proposition 7 *(Ethics,* Part II) is that 'the order and connection of ideas is the same as the order and connection of things'.

3. On several occasions in his writing, William Irwin Thompson used the metaphor of making 'archaeological excavation of consciousness'.

4. If this is not precisely what has happened, it is at least what many claim as the rational, scientific story of the human species. Alternative creation stories abound, but few omit some hypothesis of origins and progressive development involving some natural or supernatural agency.

5. William Irwin Thompson coined the term 'post-religious spirituality.' See Thompson, W.I. (2004) *Self and Society: Studies in the Evolution of Consciousness.*

6. Phillip Sherrard asked: 'If things are evolving, and if human consciousness is evolving along with everything else, where do we find a standpoint from which to understand the whole process.' In Sherrard P. (1992) *Human Image, World Image,* p. 72. Cited by Wendell Berry, *Life is a Miracle,* p. 150.

Chapter 1

1. A fuller quote shows that Hobbes was referring to individual 'minds'. He held that 'minds never meet, that ideas are never really shared and that each of us is always and finally isolated from every other individual'. Cited in Hay, S.D. (2001).

2. In his book *The Universe in a Single Atom,* the Dalai Lama discusses how Buddhist thinking, which parallels the scientific method (observation, analysis, and interpretation), nevertheless starts with subjective before objective observation, thus encompassing both inner and outer experience. Such meditative consciousness is much studied by scientists such as Richard Davidson and James Austin.

3. Curd, P. (1995) ed. *A Presocratics Reader.*

4. Kuhlewind, G. (1986) *The Logos-Structure of the World: Language as Model of Reality.*

5. Plato, *Phaedo,* and Descartes, *Meditation II.* (See References.)

6. Searle, J. (2004) *Mind: A Brief Introduction.* See also Searle, J. (1984/2003) *Minds, Brains, and Science;* Dennett, D.C. (1991) *Consciousness Explained.*

7. Demasio, R.A. (1994) *Descartes' Error: Emotion, Reason, and the Human Brain.* See also Lehrer, J. (2007) *Proust was a Neuroscientist.*

8. Plato, *Symposium,* 202a.

9. Cosmos is often used in the sense of the physical universe, but the original meaning of *kosmos* (Greek) was 'order' and could be contrasted with *khaos* (disorder). Philosophers such as Ken Wilber make a distinction between physical Cosmos and Kosmos, giving the latter a

broader meaning. See Wilber, K. (1995) *Sex ecology and spirituality: the spirit of evolution.*

10. If our self-consciousness expands 'cosmically' to include all other species in the biosphere, this may also be in our best 'environmental' interests. Some would argue that our unconscious already has this biosphere-wide reach since we have evolved as part or organ of the biosphere for 3.5 billion years.

11. This Teilhard de Chardin quote has been widely distributed on internet quotation sites.

12. *The Book of Chuang Tzu.* Translated by Martin Palmer and Elizabeth Breuilly, published by Arkana.

13. The 'Body-Mind' concept sometimes written as Bodymind has been much discussed by ancient and modern philosophers: see John Dewey for a twentieth century treatment.

14. See here Chapter 5.

15. Rudolf Steiner (1963) *The Life, Nature and Cultivation of Anthroposophy.*

16. Owen Barfield was a linguistic genius who was close friends with C.S. Lewis and J.R.R. Tolkien, with whom he and others formed the Oxford literary group known as the 'Inklings'. These famous authors, like others who know his work, regarded him as a highly original thinker.

17. Building on the prolific work of Ken Wilber, Steve McIntosh has attempted to define Integral Consciousness and Integral Philosophy. See McIntosh, S. (2007) *Integral Consciousness and the Future of Evolution,* Chapter 5.

18. Although Aristarchus knew in the third century BC that the Earth revolved around the Sun, this fact was forgotten until rediscovered by Copernicus in the sixteenth century. See Koestler, A. (1959) *The Sleepwalkers.*

19. Abbs, P. (1986) "The development of autobiography in western culture: from Augustine to Rousseau." Ph.D. thesis, University of Sussex, UK.

Chapter 2

1. The Einstein quote is reiterated several times in this book and in various sources: e.g., Einstein, A. and Calaprice, A. (2000) eds. *The Expanded Quotable Einstein.*

2. The 'Grand Synthesis' concept is attributable to Ervin Laszlo, who clearly believes in evolution, but does not support simple natural selection as the only mechanism.

3. Having made this rather stark comparison we should note that Darwin has nevertheless earned well-deserved approbation for his study of the expression of emotions in humans and animals which establishes him as one of the pioneers of modern psychology. Nevertheless, it is often forgotten that Alfred Russel Wallace was a scientific and intellectual polymath of the first order who pioneered the theory of evolution by 'natural selection' and co-authored the original report with Darwin, a year before Darwin's book *The Origin of Species.*

4. See the work of Jean Gebser reviewed here in Chapter 5.

5. Geoffroy Saint-Hilaire was a French biologist who argued, in the 1820s, against the unchanging 'fixity of species' doctrine propounded by the famous Baron Georges Cuvier. Although Saint-Hilaire was right his work was mostly forgotten until the 1990s, when he enjoyed a revival as a result of genetic studies which confirmed his profound insights into the deep relationships between all animals.

6. Fawcett, D. (1916) *The World As Imagination.*

7. Slotten, R.A. (2004) *The Heretic In Darwin's Court: The Life of Alfred Russel Wallace.*

8. Richard Dawkins' book *The Selfish Gene* has influenced a generation of biologists to apply Darwinian thinking at the molecular level.

9. Quote from Chopra, D. (2005) *Peace is the Way,* p. 9.

10. Both Mary Midgley and the Oxford theologian-biophysicist Alistair McGrath have criticized Dawkins at considerable length for his one-sided doctrines, and his strident atheistic preaching. Michael Ruse, the prolific and accomplished philosopher of biology, himself a confessed evolutionist, has also joined in the chorus of criticism denouncing Dawkins for his simplistic arguments.

11. Lynn Margulis is a famous biologist, and ex-wife of Carl Sagan, from the University of Massachusetts, Amherst. See Chapter 7.

12. Sahtouris, E. (2000) *Earthdance.*

13. The original paper on this subject is by Niles Eldredge and Stephen Jay Gould (1972): "Punctuated equilibria: an alternative to phyletic gradualism." In T.J.M. Schopf, ed., *Models in Paleobiology,* pp. 82-115.

14. William Irwin Thompson was prescient in suggesting that this would happen to weather systems and the stock market at about the present time.

15. Combs, A. (1996) *The Radiance of Being: Complexity, Chaos and the Evolution of Consciousness,* provides a good general statement about the application of chaos theory to consciousness studies.

16. Bird, R.J. (2004) *Chaos and Life: complexity and randomness in evolution and thought.* See also Mandelbrot, B.B. (1977) *The Fractal Geometry of Nature.*

17. For more on Richard Bucke, see below Chapter 3.

18. For more on William James, see Chapter 3.

19. Alan Watts is author of many books on Buddhism and ancient Eastern Wisdom.

20. For more on Alister Hardy, see Chapters 3 and 9.

21. Newberg, A., D'Aquili, E. and Rause, V. (2001) *Why God Won't Go Away: Brain Science and the Biology of Belief.*

22. Austin, J. (1998) *Zen and Brain.*

23. Russell, P. (1995/2007) *The Global Brain.*

24. Cybernetics is the study of self-regulating and self-adjusting systems with feedback loops. The word *cyber* originates from the Greek word for helmsman or pilot: the one who constantly adjusts course. For more on Teilhard de Chardin, a paleontologist priest, see Chapters 8 and 9.

25. The 'Hundredth Monkey Effect' has been much discussed by Rupert Sheldrake and others. In his 1979 book *Lifetide,* Lyall Watson cites the original study of the Japanese monkey *Macaca fuscata* by M. Kawai (1965).

26. Claxton, G. (1997) *Hare Brain, Tortoise Mind: Why Intelligence Increases When You Think Less.*

27. *What the Bleep do we know,* a 2004 film co-directed by William Arntz, Betsy Chasse and Mark Vicente, became a modest commercial success and something of a cult classic among consciousness aficionados and those interested in quantum physics.

28. The original citation for Bell's Theorem is: Bell, J.S. (1964) "On the Einstein-Podolsky-Rosen Paradox", *Physics* 1, pp. 195-200.

29. GUTs and TOEs. See for example, Ellis, J. (1986) "The superstring: theory of everything, or of nothing?" *Nature* 323: pp. 595-98.

30. See Chapter 2, note 2 above

31. Lovelock, J. (1979) *Gaia.*

32. See here Chapter 7 for more recent work by Ervin Laszlo.

33. See here Chapter 6.

34. Huxley, A. (1945) *The Perennial Philosophy.*

35. See Chapter 3 for discussion of classic child development studies.

36. Plotinus (1991) *The Enneads*. Trans. Stephen MacKenna and John Dillon.

37. Ghose, A. and McDermott, R.A. (1994) *The Essential Aurobindo*.

38. *Tao Te Ching* (Chapter 41).

39. *Tao Te Ching* (Chapter 1).

40. See above Chapter 1, note 10. If our self-consciousness expands 'cosmically' to include all other species in the biosphere this may also be in our best 'environmental' interests.

41. All quotes from in this paragraph from pages 3-6 in Yoke, H.P. (1985) *Li, Qi and Shu: An Introduction to Science and Civilization in China*. Also see Needham, J. (1954-2008) *Science and Civilization in China*.

42. Spinoza, B. de (1982) *The Ethics and Selected Letters*, trans. Samuel Shirley. (Original work published 1677).

43. Carl Jung, the famous Swiss psychologist, applied the Mandala concept to express the dimensions of self in the individual.

44. Here we refer only briefly to the well-known founders of the cited concepts/theories: the Greek Democritus of Abdera (460-370 BC) gave us the concept of atoms; the Austrian monk Gregor Mendel (1822-84) gave us the concept of genetics, and Heinrich Wilhelm Gottfried von Waldeyer-Hartz proposed the term neurons in 1891.

45. The work of Barbara McClintock was seminal in showing that genes are highly mobile and can move or 'jump' around in a process known as 'transposition' See Keller, E.F. (1983) *A Feeling for the Organism*.

46. Bortoft, H. (1996) *The Wholeness of Nature: Goethe's Way of Science*. See also Bohm, D. (1980) *Wholeness and the Implicate Order*.

47. This quote: 'A physicist is an atom's way of perceiving itself' is attributed to Niels Bohr, and has been paraphrased into: 'a physicist is the universe's way of perceiving itself'.

48. Lemkow, A.F. (1990) *The Wholeness Principle: Dynamics of Unity Within Science, Religion and Society*.

49. Steiner, H. and Gebser, J. (1962) *Anxiety: A Condition of Modern Man*.

50. In *Self and Society*, William Irwin Thompson (2004) develops the idea that dark ages and death may precede or accompany renaissance. You cannot make an omelette without breaking eggs.

Chapter 3

1. Jean Piaget was a Swiss psychologist and philosopher famous for his work on child cognitive development: see Chapter 3, note 8 below.

2. Freud. S. (1933) "New Introductory Lectures." See also Freud, S., and Strachey, J. (ed.) (1960) *The Ego and the Id*.

3. Freud and Strachey (1960).

4. Lear, J. (1990/1998) *Love And Its Place In Nature: A Philosophical Interpretation of Freudian Psychoanalysis*.

5. Lear, J. *op.cit.*, p. 176-7. This is a very similar perspective to that proposed by Owen Barfield when he said that it was only in comparatively recent times that humans could regard 'thought' as something 'within' their own heads rather than an external muse-like inspiration. See Chapters 4 and 5.

6. Freud, S. (1933) p. 80; Freud and Strachey (1960) p. 56.

7. Jung, C.G. (1931/1969) *The Structure of the Psyche.*.

8. Piaget's original work was far more detailed and voluminous. Here we select a useful summary: Piaget, J. (1976) *The Grasp Of Consciousness: Action and Concept In the Young Child*.

9. König, K. (1969) *The First Three Years of the Child*.

10. Cephalo-caudal (head to tail) and reverse caudo-cephalic (tail to head) growth dynamics were much studied in Physical Anthropology in the early twentieth century: e.g., Kingsbury, B.F. (1924).

11. We deal with the broader implications of this topic in more depth elsewhere in the book.

12. The Buddhist commentator Alan Watts says even adults divide the world, figuratively, into prickly (sharp) and squigy (soft) objects and people. There can be no doubt that we structure our language to describe all manner of hard and soft entities and tasks that confront us in life.

13. Goodenough, F.L. (1926) *Measurement of Intelligence by Drawing*. See also Coles, R. (1990) *The Spiritual Life of Children*, who reports that after asking 293 children to draw a picture of God, 87% (245) drew only the face.

14. See Goodenough (1926), Fig. 20.

15. Erickson, E.H. (1959); also Erickson, E.H. (1968) *Identity: Youth and Crisis*.

16. Endres, P. and Schad, W. (2002) *Moon Rhythms in Nature*.

17. Rembrandt, the great Dutch master, painted more than ninety self-portraits throughout his life. See Osmond, F.S. (2000) "Rembrandt's Self Portrait," *The World and I* magazine, at www.worldandi.com/specialreport/rembrandt/rembrandt.html

18. The Waldorf School system was founded by Rudolf Steiner in Stuttgart, Germany, early in the twentieth century. Since then it has spread to many parts of the world, and is regarded as a progressive and organic system that integrates education with the natural stages of childhood development. For a detailed exposition see Wilkinson, R. (1996) *The Spiritual Basis of Steiner Education*.

19. Thompson, W.I. (1973) *Passages about Earth*.

20. The Ashrama system is detailed in many sources on Hinduism.

21. *Homo Faber*: the first book with this title was written by G.N.M Tyrrell in 1951, and deals with human mental evolution, while the second book with this title, written by Max Frisch in 1957, is a novel.

22. The early onset of sexual maturity has been called the 'secular trend'. Demographic records from Europe show that the onset has occurred earlier and earlier since the 1840s. See Whincup, P.H.; J.A . Gilg, K. Odoki, S.J.C. Taylor, D.G. Cook (2001).

23. Deterioration of eyesight in late 1940s is discussed in Myers, D.G. (2007) *Psychology* (eighth edition, module 11) "Adulthood, and reflections on Developmental Issues," pp. 171-76.

24. Anecdotally, on being told how good she looked at sixty, Tina Turner is said to have replied: 'This is what the new sixty looks like.' This same theme appears to have been reiterated by Stevie Nicks, another rock goddess, in her video: 'This is what sixty looks like now.'

25. Bucke, R.M. (1901) *Cosmic Consciousness*.

26. Lockley, M.G. (2000). See also Galbraith, J. (1998).

27. Douglas-Smith, B. (1971).

28. James, W. (1908) *The Varieties of Religious Experience*.

29. Carl Jung and Max Müller (Chapters 7-9) are among notable western scholars who made serious investigations of the perennial philosophy of the East.

30. For more on the kundalini phenomenon, see Introduction, note 1, and Chapters 8 and 9.

31. Bentov, I. (1977) *Stalking the Wild Pendulum: On the Mechanics of Consciousness*. See also Sannella, L. (1987) and Galbraith, J. (1998).

32. Hardy, A. (1975) *The Biology of God*. Hardy won the Templeton Prize in 1985

just before he died of a stroke at age 89. He had already written his acceptance speech entitled: "The Significance of Religious Experience". (Occasional paper 12, Religious Experience Research Centre, Oxford University, 1997.)

33. See Chapter 1, note 1, and Hay, D. (2001) pp. 124-35.

Chapter 4

1. Santillana, G. de and von Dechend, H. (1969) *Hamlet's Mill,* Gambit, Boston. These authors argue that as early as the Neolithic (>5000 BP) humans recognized the precession of the equinoxes. This means that they knew the night sky rotated very slowly relative to earthly reference points. Thus, they recognized subtle shifts in the position of sunrise and sunset, as well as the rising and setting of the moon and other planets at key times (equinox and solstice) relative to the fixed 'background' of star constellations. These constellations included the twelve signs of the zodiac, first identified in great antiquity. This supposedly helped archaic humans appreciate the concept of cyclic time. Not only did the night sky appear to rotate about the north star every night, but there was also a more complex 'whirling' motion of the seven planets (then identified as the Sun, Moon, Mars, Mercury, Jupiter, Venus, and Saturn) which reinforced the idea that the universe was akin to a great whirlpool or grinding mill. To this day these seven planets (wanderers) correspond, respectively, to the seven days of the week, and bear the names of the best known gods of mythology.

2. Strauss, W and Howe, N. (1997) *The Fourth Turning.*

3. See Santillana, G. de and von Dechend, H. (1969) and Chapter 4, note 1 above, for further details of how complex planetary motions and conjunctions with astronomical constellations are reported and symbolized in ancient literature. This complex process of integrating complex astronomical observations with hard-to-decipher ancient literature, involves strenuous scholarship and ambiguous results that lead to controversial interpretations.

4. Bronowski, J. (1965) *William Blake and the Age of Revolution.* In this excellent little book, Jacob Bronowski explores Blake's political and emotional response to the societal upheaval generated by the Industrial Revolution. The exposition shows that Blake understood the deeper spiritual tensions inherent in human nature and especially between humankind's rational and non-rational faculties.

5. Armstrong, K. (2006) *The Great Transformation.*

6. Keck, L.R. (2000) *The Sacred Quest: The Evolution and Future of the Human Soul.*

7. Eisler, R. (1987) *The Chalice and the Blade.*

8. Berman, M. (1989) *Coming To Our Senses: Body And Spirit in the Hidden History of the West.*

9. The four preceding quotes are taken from Berman (1989).

10. This famous quote is attributed to Charles Péguy (1873-1914). Quote from p. 21 of Gilbert, S.R. (2000) *The Prophetic Imperative: Social Gospel in Theory and Practice.*

11. For more on Owen Barfield, see Chapter 5.

12.. Plotinus is often labelled as a Neo-Platonist and mystic. Although he lived two centuries after Christ and sometime after the appearance of the influential mystic writings of John and Paul, he was not a Christian. However, he is often regarded, especially in Roman circles, as the single most influential spiritual figure of this period leading up to the Council of Nicea in 325. See Cheney, S. (1945) *Men who Have Walked with God.*

13. LeRoy Ladurie, E. (1980) *Montaillou: Cathars and Catholics in a French Village*, 1294-1324.

14. Cited in Meister Eckhart's *Sermons* (Sermon IV: True Hearing) translated by C. Field (1909).

15. Thompson, W.I. (1981) *Time Falling Bodies Take To Light.*

16. Brooke, J.H. (1991) *Science and Religion.*

17. Brooke, J.H. *op.cit.*

18. One dimension of alchemy was related to the quest for transformation (of matter and the individual soul).

19. The Knights Templar brotherhood was founded by the mystic St Bernard of Clairvaux (1090-1153) in what has been called the Age of Renewal.

20. Helena Blavatsky (1831-91) is famous for founding the esoteric spiritual movement known as Theosophy, and for her many writings including her magnum opus, *The Secret Doctrine* (1888).

21. The famous quote - 'those who do not learn the lessons of history are condemned to repeat them' - is attributed to the Spanish philosopher George Santayana (1863-1952).

22. Yutang, L. (1935) *My Country My People.*

23. Kuhn, T. (1962) *The Structure of Scientific Revolutions.*

24. Edelglass, S. (2006) *The Physics of Human Experience;* Dobbs, B.J.T (1975) *The Foundations of Newton's Alchemy.*

25. Hapgood, C. (1964) *Maps of the Ancient Sea Kings: Evidence of Advanced Civilization in the Ice Age.* Some critics say the southern continent depicted on these maps is Australia, even though it is centred on the South Pole.

26. Toumlin, S. and Goodfield, J. (1967) *The Discovery Of Time.*

27. *Ibid.* p. 25.

28. Steiner, R. (1971) *Cosmic Memory.*

29. Toumlin, S. and Goodfield, J. (1967) p. 27. This sentiment originates from Plato's *Phaedrus*. Santillana and von Dechend (1969, p. 348) cite the translation as follows: 'this invention [writing] will produce forgetfulness in the minds of those who learn to use it, because they will not practise their memory.'

30. Cited in Shepherd, A.P. (1983) *Scientist Of The Invisible: Rudolf Steiner.*

31. Toumlin, S. and Goodfield, J. (1967) p. 38.

32. Manchester, W. (1992) *A World Lit Only By Fire.*

33. See Chapter 4, note 15 above.

34. See p. 82 in Blaxland de Lange, S. (2006) *Owen Barfield: Romanticism Come of Age - A Biography.*

35. Eddington, A. (1964) *The Nature Of The Physical World.*

36. Conan Doyle, A. (1921) *The Lost World;* Verne, J. (1870) *Vingt Mille Lieues Sous Les Mers.* First published in English in 1872, as *20,000 leagues Under the Sea.*

37. Swift, J. (1726) *Gulliver's Travels;* Abbott, E. (1884) *Flatland: A Romance Of Many Dimensions;* Carroll, L. (1866) *Alice in Wonderland.*

38. Bell, E.T. (1951) *Mathematics: Queen and Servant of Science.*

39. Einstein, A. (1961) *Relativity.*

40. Eddington (see note 35 above) first coined the term 'Time's arrow'. For further details see Blum, H.F. (1951) *Time's Arrow and Evolution.*

41. Bergson, H. (1988) *Matter and Memory,* p.186 (italics original).

42. Priestly, J.B. (1937) *Midnight on the Desert.*

43. Dunne J.W. (1927) *An Experiment with Time.*

44. For more on the collective unconscious, see Carl Jung (1981) *The Archetypes and the Collective Unconscious.* Also see Jung, C.G. (1966/1977) *Two Essays on Analytical Psychology.* For Steiner's 'group soul', see Steiner, R. (1999) *A Psychology of Body, Soul and Spirit.* Trans. Marjorie Spock.

45. Sheldrake, R. (1981) *A New Science Of Life.*

46. Homeostatis is the self-regulation of an organic system. It literally means the 'same standing' or the same position and implies regulation or equilibrium.

47. Wilson, E.O. (1999) *Consilience.*

48. Stephen J. Gould described science and religion as two 'Nonoverlapping Magisteria.' A much debated term abbreviated to NOMA. See Gould, S.J. (1997).

Chapter 5

1. See Chapter 2, note 1 above.

2. This Owen Barfield quote (see www. netfuture.org/fdnc/appa.html) originates from his books *Speaker's Meaning,* and *Saving the Appearances.*

3. The field of 'cognitive archaeology' as applied to the interpretation of 'Old Stone Age' tools, that is, Lower Paleolithic Acheulian hand axes, has generated much interest in the last decade or two. The Feliks quote comes from Feliks, J. (2008) "Phi in the Acheulian", in Bednarik, R.G. and Hodgson, D. (eds.) *Pleistocene Paleoart of the World,* p. 11-31, British Archaeological Reports International Series 1804, Oxford. For further discussion see Mithen, S. (2003) "Handaxes: the first aesthetic artifacts",in Eckhart, V. and Grammer, K. (eds.) *Evolutionary Aesthetics,* pp. 261-75, Springer Verlag, Berlin. Kohn, M. and Mithen, S.J. (1999) "Handaxes: products of sexual selection", *Antiquity,* 73: pp. 518-26. See also Mithen, S.J. (1996) *The Prehistory of the Mind,* Thames and Hudson, London.

4. Clottes, J. and Lewis Williams, D. (1998) *The Shamans of Prehistory: Trance and Magic in the Painted Caves.*

5. Pollack, J.H. (1965) *Croiset the Clairvoyant.*

6. Gebser, J. (1949/1986) *The Ever-Present Origin.* In the original German title

Ursprung und Gegenwart, Ursprung means 'primary leap' and *Gegenwart* means 'present'. In German, as in English, a distinction is made between consciousness and conscience that does not exist in many other languages either today or in the past.

7. Gebser (1986) p. 42. Here the adjective 'wakeful' appears to imply self-consciousness.

8. Gebser (1986) p 38.

9. Gebser (1986) p. 7.

10. Gebser (1986) p 43.

11. Gebser (1986) p. 44.

12. Gebser (1986) p. 45.

13. Mircea Eliade.

14. Teleology or entelechy, from Greek *teleos/telos* meaning 'from afar', refers to the notion of cosmic design, purpose, goal or completion. In as much as humans have purpose, and future goals, human psychology is highly teleological. Whether organic evolution is purposeful, as in the seed's goal to produce a plant and flower, is open to scientific and philosophical debate.

15. Gebser (1986) p. 46.

16. Frobenius, L. (1905) *Kulturgeschichte Afrikas,* Phaidon, Vienna, cited in Combs, A. (2002) *The Radiance of Being* (second edition).

17. 'Sorcery Killers' Jailed. *Post Courier,* Papua New Guinea, March 4 1992, p. 4. Today we tend to call these causes 'supernatural', which simply indicates that we have devised a separate label for natural phenomena we don't understand.

18. See Chapter 5, notes 4 and 13. Shamanism has been referred to as an archaic, universal 'religion', older and more widespread than any other for which records exist. The Shaman is able to transform his/her consciousness to enter another realm of consciousness - to enter state of trance or 'ecstasy', to literally 'stand outside' *(ex stasis)* him/herself.

19. 'Macrocosmic harmony' does not necessarily imply some ideal utopia. Unconscious harmony with nature involves birth, death (killing) and the so-called 'struggle for survival'.

20. Long, B. (1984) *The Origins of Man and the Universe: The Myth that Came to Life.* Barry Long argues that all our mythology of gods, demigods and demons is in fact the manifestation of actual psychic experience that arose was we became self-conscious. The gods represent real psychic experiences, much like dreams, that result from becoming aware of our physical and emotional/psychic sensory experience and associating it with our physical bodies for the first time.

21. Frobenius, L. (1905) p. 48.

22. The well known placebo effect proves that the mind can play a 'magic' trick by curing the body without any physical intervention

23. Elwood Babbitt, an authentic medium, was analyzed by a psychologist and found not to be schizophrenic. He could, at will go into a trance and receive messages from disincarnate spirits, who provided information impossible for him to have known by other means. His case was documented by Hapgood, C. (1975) *Voices of Spirit.* Similar cases have been documented by eminent psychiatrists, for instance, Weiss, B.L. (1988) *Many Lives, Many Masters.*

24. Van der Post, L. (1958) *The Lost World of the Kalahari.*

25. Sheldrake, R. (1999) *Dogs That Know When Their Owners Are Coming Home.*

26. Julian Jaynes (1976) *The Origin of Consciousness in the Breakdown of the Bicameral Mind.*

27. Carl Jung on synchronicity and coincidence, in Carl Jung (1972) *Synchronicity - An Acausal Connecting Principle.*

28. Gebser (1986) p. 145.

29. See p. 189 in Nitobé, I. (1998) *Bushido, the Soul of Japan.*

30. Gebser (1986) p. 61.

31. Gebser (1986) p. 66.

32. Holbrook, B. (1981) *The Stone Monkey.*

33. Gebser (1986) pp. 61-73.

34. See Chapter 1, note 12.

35. Campbell, J. (1949) *The Hero With a Thousand Faces,* Bollingen, New York.

36. Gebser (1986) p. 70.

37. Gebser (1986) p. 67.

38. Barfield, O. (1967) *Speaker's Meaning.* This is one of many Barfield titles that are worth reading. Incidentally, no one really knows when language emerged. In the 1860s and 1870s, speculation on the subject was so rife that, with in the space of a few years, both the Linguistic Society of Paris and the London Philological Society banned any further essay contributions on the subject.

39. Jung, C. (1961) p. 248.

40. Marshack, A. (1972) p. 67-68.

41. See Lavie, P. (1996) *The Enchanted World of Sleep,* p. 244.

42. Gebser (1986) p. 41. Also p. 12-13.

43. When first 'experienced' by the early Greeks, it was as if the gods had endowed man with a new faculty - 'A gift from on high' - thought, reason, logos, the ability to grasp knowledge or know something.

44. Gebser (1986) p. 77.

45. Socrates insisted that people use their new faculty of thought to reflect and examine their own existence: to be thoughtful. Conversely, the term 'thoughtless' has a negative connotation today. But many of our ancestors were thoughtless in quite another sense. They did could not and did not yet think discursively.

46. Gebser (1986) p. 77.

47. Gebser (1986) p. 97.

48. Gebser (1986) p. 99.

49. All quotes from Gebser (1986) p. 98-99.

50. Feuerstein, G. (1987) *Structures of Consciousness: the Genius of Jean Gebser - an Introduction and Critique*. In this work, Feuerstein refers to Gebser's autobiographical reference to 'the sleeping years' (in his early twenties), which we may contrast with his *satori* experience in his mid-fifties. Conscious experience of 'intensified,' states of consciousness has been described by many authors as 'spiritual experience' (Chapter 3) or higher or expanded consciousness (Chapter 8). Some such as Barry Long (1984) consider this a conscious reconnection with the unconscious psychic web of the biosphere. Fawcett refers to Universal Psychical Life. We warn that different descriptions by different authors may refer to different states of consciousness. However, while no-one is objectively qualified to categorize experiential states without ambiguity, certain consistent patterns of experience are noted, particularly regarding light and illumination (see Chapter 3).

51. For quotes in this paragraph, see Feuerstein 1987, pp. 7-16 and Glossary, pp. 211-21.

52. Gebser (1986). p. 58, states that integration is not the same as 'synthesis'.

53. Stephen Edelglass makes the interesting observation that the wave-particle duality problem is actually an artifact of our rationalization of the world as consisting of process and object respectively, which in turn is related to our artificial division of the world into subjective and objective phenomena. See Edelglass, S. (2006) *The Physics of Human Experience*.

54. Combs, A. (1996).

55. Thompson, W.I. (1996) *Coming into Being*.

56. Barfield, O. (1988) *Saving the Appearances: Studies in Idolatry* (second ed.).

57. Ken Wilber has presented his four quadrant scheme in many of his publications most notably in: Wilber, K. (1995) *Sex, ecology and spirituality: the spirit of evolution*; Wilber, K. (1998) *The Marriage of Sense and Soul,* and in a journal article, Wilber 1997.

58. Wilber, K. (1997).

59. Combs, A. (1996). See also McIntosh, S. (2007).

60. Wilber has refined and repackaged his own work several times, creating his self-referential labels Wilber I-Wilber IV. Combs has also adopted these labels in his descriptions and re-analysis of Wilber's work.

61. Clare W. Graves.

62. McIntosh (2007) p. 32-33.

63. See Chapter 5, note 52 above.

64. McIntosh (2007) pp. 160-94.

65. Vico, G. (1725/1994) *The Life of Giambattista Vico written by himself.*

66. McCabe, J. *Biographical Dictionary of Modern Rationalists.* (Cited in Vico 1994.)

Chapter 6

1. There are many studies referring to a conscious universe. The following five are representative: Hubbard, B.M. (1998) *Conscious Evolution*; Edelman, G.M. and Tononi, G. (2000) *A Universe of Consciousness*; Kafatos, M. and Nadeau, R. (1990) *The Conscious Universe: Part and Whole in Modern Physical Theory*; Goswami, A. (1995) *The Self Aware Universe: How Consciousness Creates the Material World;* Radin, D.I (1997) *The Conscious Universe.*

2. Many neuroscientists are still trying to find the neural correlates of consciousness in very specific locations in the brain.

3. Wilber, K. (1977) *The Spectrum of Consciousness*.

4. Lakoff, G. and Johnson, M. (1999) *Philosophy in the Flesh*.The subtitle of this book 'the embodied mind and its

challenge to western thought' clearly suggests that the problem of body-mind integration is far from resolved in western thought.

5. Lovelock, J. (1979).

6. Habermas, J. (1984-85) *The Theory Of Communicative Action,* (2 vols) trans. T. McCarthy.

7. Ho, M-W. (1996).

8. The choice of simple over complex explanations is sometimes referred to as 'Occam's Razor'.

9. Few scientists doubt that the biosphere is a system in homeostasis. But semantic and conceptual problems and debates arise when the system is described as sentient or conscious. See Chapter 6, note 1 above.

10. Weiss, P. (1973) *The Science of Life.*.

11. Koestler, A. (1969).

12. Talk of forces of disintegration and integration, create the danger of value judgments. If integration, integral consciousness and integrated societies, are better than disintegration, we should at least be aware of using the mental, measuring perspective. Paradoxically, in order to integrate oneself, with the other one must disintegrate boundaries (and the mental-intellectual egos).

13. This Cambridge USA versus Cambridge UK played out in two books. The first by Gould advocated that the hypothetical replay of life scenario would produce new results. The reply by Conway-Morris, suggested that the reply would produce similar lines of evolution to those we have actually witnessed. Gould, S.J. (1989) *Wonderful Life,* and Conway- Morris, S. (1998) *The Crucible of Creation.*

14. A substantial literature exists on the longevity of reef systems in geological time.

15. Many authors have noted that the sexual reproductive, or cloning mode of reproduction by early unicellular species was a type of immortality.

16. Verhulst, J. (2003) *Developmental Dynamics in Humans and other Primates: Discovering Evolutionary Principles Through Comparative Morphology.* Provides much detailed information on primate longevity and breath frequency in mammals.

17. Primates and humans have above average longevity for their bodyweights. This seems to be correlated with brain size. See previous note 16.

18. Rohen, J.W. (2007) *Functional Morphology. The Dynamic Wholeness of the Human Organism*, provides a masterful, in-depth look at Chronobiology. Also, Endres, P. and Schad, W. (2002) *Moon Rhythms in Nature,* provide a fascinating account of how breeding cycles of hundreds of different species are synchronized to many subtle sidereal cycles.

19. McNamara, K. (1997) *Shapes of Time: the Evolution of Growth and Development.*

20. It has been suggested that fish (the first true vertebrates) evolved from sea squirts (sedentary jelly-like blobs classified as primitive chordates) by the heterochronic process know as 'neotony'. The juvenile sea squirt has small fish like larvae, that swim freely. If they retained this juvenile 'neotonous' condition as breeding adults, without metamorphosing into sedentary, jelly-like blobs, they could have given given rise to the free swimming fish.

21. Gould, S.J. (1977b) *Ontogeny and Phylogeny.*

22. Haeckel, E. (1866).

23. Goethe predicted the existence of the human intermaxilliary bone, even though anatomists had previously denied its existence and interpreted its 'absence' as evidence of the difference between humans and animals. He found it as a very small sliver in juveniles. Goethe reasoned holistically, like Saint-Hilaire (see Chapter 6, note 30 below), that all

animals (mammals) had similar basic designs, and that the absence of a feature meant only that it was underdeveloped relative to other organs. So a bone that was large in one species might be small, or even latently hidden as cartilage or tissue in another.

24. Goethe, J.W. (1795).

25. See Chapter 6, note 19 above.

26. Schad, W. (1977).

27. Synergy is the combination of two or more entities to produce a whole that is greater than the sum of the parts. Thus, Verhulst defined 'synergistic composition' as an integration (synergy) of two characteristics that are prominent in human development: the Type I trends towards fetalization and retardation (as in nakedness and late eruption of teeth) which lasts longer in early (juvenile) human development than in other animals, and Type II trends which appear late in human development (such as our long post-reproductive phase of life or growth of neocortex and prefrontal lobes). Together these dynamics synergistically allow new potentials to develop, including speech. Type I retardation (not rushing to become specialized) allows individuals to exploit prolonged, late developing Type II capacities.

28. Kingsbury, B.F. (1924).

29. Homeobox and their associated Hox genes have been very much discussed in recent years and show interesting patterns of organization: 'what is ... surprising is that there is a direct match between the front and back of both the gene complex and the actual body of the animal. In other words the regions of the fly head are coded for by the genes at the front end of the complex, and so on back through the tail end of the abdomen...' 'What is even more remarkable is that when the Hox complex of the mouse is compared it proves to have the same basic arrangement as is found in [the fly] Drosophila' (Conway-Morris 1998, p. 148). This shows how much different groups (and modes or levels of organization) have in common.

30. Saint-Hilaire, G. (1822) "Considérations générales sur la vertèbre," *Memoir du Musée National de l'Histoire Naturelle* 9: pp. 89-114.

31. Le Guyader, H. (2004) *Geoffroy Saint-Hilaire: A Visionary Naturalist,* trans. M. Grene.

32. Gould, S.J. (1985) "Geoffroy and the Homeobox," *Natural History,* 94, pp.12-23.

33. Badyaev, A.V. (2002).

34. Schad, W. (1977).

35. The abdominal muscles, if well developed, show the segmented 'six-pack' pattern coveted by body builders.

36. Riegner, M. (1985).

37. Lockley, M.G. (1999) *The Eternal Trail: a Tracker Looks at Evolution.* See also Lockley, M.G. (2007b).

38. Riegner, M. (2008).

39. See note 37 above.

40. Portmann, A. (1964) *New Paths in Biology.* The neopallium index is the ratio between the posterior brain stem and anterior neocortex or primitive and advanced/derived parts of the brain. Smaller animals (with smaller indices) incline to be too open and suffer sensory overload. Larger animals incline to be closed, oblivious or self-contained.

41. Shelden, W.H. (1944) *The Varieties of Temperament.* See also Jung, C.G. (1976) *Psychological Types.*

42. Holbrook, B. (1981) *The Stone Monkey.*

43. Chesterton, G.K. (1919) p. 168 of sixth edition.

44. Rosslenbroich, B. (2009) "The theory of increased autonomy in evolution: a proposal for understanding macroevolutionary innovations," *Biology and Philosophy, DOI* 10.1007/s10539-009-9167-9.

45. It is also well known that during evolution plants have progressively internalized their reproductive system. They began with external spores, then developed 'naked' seeds or cones, and then finally 'covered' seeds associated with fruit and flowers.

46. Hopson, J.A. (1980). See also Rightmire, G.P. (2004).

Chapter 7

1. In his synthetic studies Ken Wilber claims to have reviewed much of the relevant literature in twelve disciplines pertinent to consciousness studies: e.g. Wilber, K. (1997).

2. Freeman, A. (2003) *Consciousness: A Guide to the Debates*. For a review of this book, see Lockley M.G. (2007a).

3. Freeman, A. (2003) p. 50.

4. There is a considerable literature on Evolutionary Psychology, derived in part from Sociobiology. See, for instance, Crawford, C. and Krebs D. (2008) *Foundations of Evolutionary Psychology*; also Buss, D.M. (2004) *Evolutionary Psychology: The New Science Of The Mind*. Sociobiology was put on the map by Harvard professor E.O. Wilson in an attempt to explain human nature in Darwinian terms tracing gender differences, aggression, altruism, religion and morality back to our Ice Age ancestry. The theory is highly controversial and was criticized by Wilson's own Harvard colleagues including Stephen J. Gould. See Wilson, E.O. (1975) *Sociobiology, The New Synthesis*.

5. The term 'survival of the fittest' was coined by Herbert Spencer and adopted by Darwin, at the suggestion of Alfred Russel Wallace as an alternative to the term 'natural selection'. Wallace felt 'natural selection' connoted too much God involvement.

6. Dawkins, R. (1976) *The Selfish Gene*.

7. David Loye shows that Darwin used the word love almost one hundred times in *The Descent of Man*, and concluded that humans, unlike animals, were capable of living by the golden rule. Loye, D. (2000) *Darwin's Lost Theory of Love*.

8. Margulis, L. (1970) *The Origin of Eukaryotic Cells*.

9. Margulis' cooperative scenario is supported by Elisabet Sahtouris, another female biologist of her generation. See Sahtouris, E. (2000) *Earthdance*.

10. Kreb's cycle also known as the Citric Acid cycle is well known to biologists and is explained in almost any standard biology or physiology text.

11. Mishlove, J. (1975) *The Roots of Consciousness*.

12. The yogic concept of the chakras, defined in the text, has become very popular in western culture in recent years. A Google search (September 2009) produced 2.5 million hits. For more information, see Chapter 9.

13. Zhang, C. (2003).

14. G.K. Chesterton writes as follows: 'If we want to uproot inherent cruelties or lift up lost populations we cannot do it with a scientific theory that matter precedes mind; we can do it with the supernatural theory that mind precedes matter. If we wish specially to awaken people to social vigilance and tireless pursuit of practice, we cannot help it much by insisting on the Immanent God and the Inner Light: for these are at best reasons for contentment; we can help it much by insisting on the transcendent God and the flying and escaping gleam; for that means divine discontent.' (1919, p. 258.)

15. Vitaliano, G. (2000).

16. MacLean, Paul D. (1990) *The Triune Brain In Evolution: Role In Paleocerebral Functions*.

17. McIntosh , S. (2007, p. 74) states that: 'The integral worldview ... fully recognizes the legitimacy and evolutionary

necessity of all previous stages of development. Integral consciousness thus grows up by reaching down ... *the degree of our transcendence is measured by the scope of our inclusion.'* (original italics)

18. Wilber, K. (1979) *No Boundary*; see also Ho, M-W. (1998) *The Rainbow and the Worm: The Physics of Organisms.*

19. Ho, M-W. (1996).

20. Penrose, R. (1989) *The Emperor's New Mind..*

21. Freeman, A. (2003) p. 174.

22. 'Wave function collapse' is the term given to the theory that the position of minute quantum 'particles' (electrons/photons) cannot be known until they are observed. This relates to Heisenberg's 'uncertainty principle' which holds that one cannot know both position and velocity. Only the act of observation fixes the position of the quantum. 'Action at a distance' may also be referred to as the 'nonlocality' phenomenon. This can be illustrated by imagining two photons that share the same wave function, traveling in opposite directions at the speed of light. If one is observed and collapses, the other, its twin, does the same even though communication between the two in theory would have to be at twice the speed of light. Freeman, A. *op.cit.* p.175.

23. Freeman, A. (2003) p. 169

24. Penrose (1989) p. 433 There are several versions of the Anthropic Principle.

25. In thermodynamics, entropy is defined by the Oxford Dictionary as 'a measure of the unavailability of a system's thermal energy for conversion into mechanical work.' More popularly it is the 'running down' of energy in a system when there is no new energy input. Thus hot coffee always cools down to room temperature, and never gets warmer.

26. Teleology is defined as the doctrine of final causes. The view that developments are due to the purpose or design that will be fulfilled by them. Human activity is highly teleological, but there is debate as to whether this is the case with non-human systems such as the universe.

27. Emerson, R.W. *Essays and Poems.*

28. Verhulst, J. (2003).

29. Although humans mature slowly in comparison with other primates, historical records show that they have been reaching sexual maturity at progressively younger ages. This phenomenon is discussed by S.J. Gould in a chapter entitled *Human babies as embryos,* in his *Ever Since Darwin* (1977a).

30. Rohen, J.W. (2007) *Functional Morphology.*

31. Verhulst (2003) p. 344.

32. Verhulst (2003) p. 112.

33. Verhulst (2003) p. 112.

34. Thompson, W.I. (2004).

35. Louis Bolk was an innovative Dutch biologist who published mainly in German. His work is cited mostly by those like Verhulst, Schad and Gould who take a serious interest in Goethean science, heterochrony and developmental dynamics. For instance, Gould, S.J. (1977b) *Ontogeny and Phylogeny.*

36. Verhulst (2003) p. 352.

37. Verhulst (2003) p. 354. Here Verhulst cites Bolk (1918) *Hersenen en Cultur.*

38. Verhulst (2003) p. 360-61.

39. Schad, W. "Scientific Thinking as an Approach to the Etheric", in Bockemühl, J. (1985) ed. *Toward a Phenomenology of the Etheric World,* pp.163-98.

40. Whitehead cited in Du Noüy (1947) *Human Destiny*, p. 43.

41. Gould, S.J. (2002) *The Structure of Evolutionary Theory.*

42. The idea of animals as automata can be traced back to the thirteenth century, but it was made famous by René Descartes who argued that animals were automata because, unlike humans, they lacked souls.

43. Spetner, L. (1997) *Not by Chance*, pp. 51 and 199.

44. Goldstein, K. (1963) *The Organism: A Holistic Approach to Biology*.

45. The idea of passing on traits acquired during an individual life time is often labeled as 'Lamarckism' after Jean-Baptiste Lamarck (1744-1829). Although mostly abandoned in favor of Darwinism, Lamarckism applies to cultural evolution and is being reconsidered in many instances by evolutionary geneticists, for instance Holdredge, C. (1996) *Genetics and the Manipulation of Life*.

46. Lockley, M.G. (2005).

47. Laszlo, E. (1987) *Evolution: the Grand Synthesis*.

48. Laszlo, E. (2004) *Science and the Akashic Field: An Integral Theory of Everything*.

49. *op.cit.*

50. The nineteenth century concept of 'luminiferous aether' was proposed to explain how light was transported.

51. Vladimir Vernadsky's classic 1945 paper "The Biosphere and Noosphere", *American Scientist* 33: pp. 1-12, first defined these now familiar terms, and gave Teilhard de Chardin credit for inventing the term 'noosphere'. See also Russell, P. (1983/2007) *The Global Brain*.

Chapter 8

1. This quote, attributed to Heraclitus in *On the Universe* has been translated as 'You could not step twice into the same river, for other waters are ever flowing on to you'. A similar rendition: 'We step into and we do not step into the same rivers. We are and we are not' is quoted in Curd, P. (1995) ed. *A Presocratics Reader*.

2. Hume, D. (1779) *Dialogues Concerning Natural Religion*. (Originally published three years after Hume's death without the publisher's name).

3. Wilson, E.O. (1999) *Consilience*.

4. Snow, C.P. (1959) *The Two Cultures and Scientific Revolution*. The 'Third Culture' concept was proposed by John Brockman in his book of that title: Brockman, J. (1995) *The Third Culture: Beyond the Scientific Revolution*. Many of the forward-looking scientists and writers cited in Brockman's book are given as examples of Third Culture representatives.

5. Steiner, H. and Gebser, J. (1962) *Anxiety: A Condition of Modern Man*.

6. Ruse, M. (2008) *Evolution and Religion: A Dialog*, p. 127-28.

7. Barfield used positivism in much the same way that the term natural materialism is used. His biographer Blaxland de Llange devotes a chapter to Barfield's ideas on 'Positivism and its Residues', or what he called the 'residue of unresolved positivism'. (See this chapter, note 32 below.) Barfield even used the abbreviation 'RUP'. See Sugerman, S. ed. (1976) *Evolution of Consciousness*.

8. Gödel's theorem comes to a similar conclusion: namely, the axioms of any system cannot be defined by that system, but rather must be defined by a higher system.

9. Harman, W. (1988) Global Mind Change.

10. Chesterton, G.K. (1919) p. 158.

11. Teilhard de Chardin, P. The most accessible scientific reference to Teilhard's coining of the term "noosphere" is Vladimir Vernadsky's 1945 paper "The Biosphere and Noosphere." (See Chapter 7, note 51 above.)

12. Teilhard de Chardin, P. (1964) *The Future of Man*, p. 63. (See discussion below: note 60.)

13. Milner, R. (1990) p. 338.

14. Teilhard de Chardin, P. (1964) p. 57.

15. Teilhard de Chardin (1964) p. 63.

16. Jennings, H.S. (1930) The *Biological Basis of Human Nature*, p. 364.

17. Jennings, H.S. (1930) p. 369.

18. Long (1984) p. 15.

19. Long (1984) p. 17-18.

20. Quote from p. 44 in Vitaliano, G. (2000).

21. This Carl Jung quote is from Radin, D. (1997) *The Conscious Universe*, p. 290.

22. Long (1984) p. 27-28.

23. Long (1984) p. 29.

24. Long (1984) p. 34-37.

25. Ruse (2008) p. 126.

26. Long (1984) p. 6-7.

27. One Gary Larson cartoon with the caption "Quick! Anthropologists!" shows indigenous people scrambling to hide.

28. Eliade, M. (1964) *Shamanism*.

29. Quotes from Aczel, A.D. (2009) *The Cave and the Cathedral*, p. 120-21.

30. Gleiser, M. (1997) *The Dancing Universe: From Creation Myths to the Big Bang*. Gleiser points out that there are two polar cosmogonic myths: one speaks of an eternal universe, with no beginning; the other postulates an abrupt beginning.

31. Muller, F.M. (1873) *Lectures on the Origin and Growth of Religion as Illustrated by the Religions of India*.

32. Blaxland de Lange, S. (2006) p. 93.

33. Hardy, A. (1975).

34. Pals, D.L. (1996) *Seven Theories of Religion*.

35. Krishna, G. (1972) *The Secret of Yoga*.

36. Muller (1873) p. 58.

37. Barfield, O. (1967) *Speaker's Meaning*. Other significant books by Barfield are: (1926/1988) *History in English Words*; (1965/1988) *Saving the Appearances: A Study in Idolatry*.

38. Cassirer, E. (1946) *Language and Myth*, see translator's preface, pp. viii-x.

39. Cassirer, E. (1946) p. 3.

40. Cassirer, E. (1946) p. 12.

41. Usener, Hermann (1896) *Götternamen: Versuch einer Lehre von der religiösen Begriffsbildung*. Cited in Cassirer (1946, p. 15) with the following translation of the subtitle: 'An Essay towards a Science of Religious Conception'.

42. Cassirer, E. (1946) pp. 31, 35, 38, 48.

43. Norman Mailer recounts this story in the documentary film: *When we were Kings*.

44. Cassirer, E. (1946) p. 62-72.

45. Durkheim, E. (1951) *Elementary Forms of Religious Life*, p. 416.

46. Cassirer, E. (1946) p. 78.

47. Cassirer, E. (1946) p. 81.

48. Cassirer, E. (1946) Chapter 6.

49. Bortoft, H. (1996) *The Wholeness of Nature*, p. 310.

50. There are many reports of feral children, One of the most beautifully and sensitively written is Armande, J-C. (1974) *Gazelle Boy*, Bodley Head, London.

51. Grotstein, J. S. (1997).

52. Girard, R. (1986) *The Scapegoat*.

53. Fromm, E. (1941) *Fear of Freedom*, published in the USA as *Escape from Freedom*.

54. Genesis, Chapter 22.

55. Young, D. (1991) *Origins of the Sacred*, pp. 226 and 229.

56. Fadiman, A. (1997) *The Spirit Catches You and You Fall Down*.

57. Bucke, R.M. (1901) *Cosmic Consciousness*.

58. James, W. (1908) *The Varieties of Religious Experience*.

59. Steinbock, A.J. (2007) *Phenomenology and Mysticism*. See also James, W. (1908); Otto, R. (1969) *The Idea of the Holy*; Underhill, E. (1911) *Mysticism*; Happold, F.C. (1963) *Mysticism: A Study and an Anthology*. See also Abhayananda, S. (1996) *History of Mysticism: the Unchanging Testament*.

60. Medawar, P.B. in his (1996) *The Strange Case of the Spotted Mice, and Other Classic Essays on Science*, levels criticism at Teilhard de Chardin's (1959) *The Phenomenon of Man*. Medawar's book *The Future of Man* appeared in the same year as the translation of Teilhard's book with the same title (1964).

61. Forman, R.K.C. (1998). See also Hunt, H. (2006).

62. Many researchers already cited directly and indirectly endorse mysticism as an experiential path to intensified or higher, integral consciousness. These include Wilber, Combs, McIntosh (see Chapter 5).

63. Lodge, R.C. (1956) *The Philosophy of Plato*.

64. Steiner, R. (1961) Christianity as Mystical Fact.

65. Cassirer, E. (1944) *An Essay on Man*.

66. Ravindra, R. (1998) *Christ the Yogi: A Hindu Reflection on the Gospel of John*. Revised (2004) as *The Gospel of John in the Light of Indian Mysticism*.

67. Pagels, E. (1979) *The Gnostic Gospels*.

68. Plotinus, *Ennead* vi. 9. Cited in Underhill, E. (1995) *Mysticism*, p. 372-73 (originally published in 1911, by Methuen).

69. Cahill, T. (1996) *How The Irish Saved Civilization*.

70. The *Life of Anthony* was written by Athanasius (ad 296-373).

71. *The Doctrine of Free Spirit*. Attributed to the free-thinking Paris professor, Amaury of Bena, gave rise to the sect known as the Amaurians, and the 'Brethren of the Free Spirit'. Pope Innocent III summoned Amaury to appear at Rome in the early thirteenth century and condemned his views. This was one of many independent movements, like the Cathars, that heralded reformation thinking

72. See www.eckhartsociety.org for various Meister Eckhart quotes.

73. Desiderius Erasmus (1466-1536) is considered the 'father' or 'Prince of Humanism'.

74. Chesterton, G.K. (1957) *St Francis of Assisi*. The five sequential quotes in the preceding paragraph come from pp. 57, 70 42, 74 and 88.

75. Bucke, R.M. (1901) *Cosmic Consciousness*, p. 9-10.

76. Krishna, G. (1975) *The Awakening of the Kundalini*, p. viii.

77. Report from the records of the Religious Experiences Research Centre, University of Wales. On file in MS by M. Lockley.

Chapter 9

1. Saul-Paul Sirag, in Mishlove, J. (1975) p. 327.

2. It is well known that all Teilhard de Chardin's major spiritual and philosophic writings were censored by the church and so only published posthumously.

3. Stace, W.T. (1960) *The Teachings of the Mystics*.

4. Ruthven, M. (1997) *Islam: A Very Short Introduction*.

5. The following citations are taken from various sources: see Chapter 8, note 59 and this chapter, note 3.

6. Rumi, J-D. (2001) *The Selected Poems of Rumi*, trans. R.A. Nicholson.

7. Polanyi, M. (1964) *Science, Faith and Society*, p. 17.

8. Polanyi, M. (1964) p. 35.

9. Rumi, J-D. (2001).

10. Thompson, W.I. (1981) *Time Falling Bodies Take to Light*, p. 87.

11. Barfield's final participation language is often difficult to follow (see Chapter 8, notes 7 and 37 for references).

12. Einstein stated that: 'The most beautiful emotion we can experience is the mystical. It is the power of all true art and science. ... To know that what is impenetrable to us really exists, manifesting itself as the highest wisdom and the most radiant beauty, which our dull faculties can comprehend only in their most primitive forms — this knowledge, this feeling, is at the center of true religiousness. In this sense, and in this sense only, I belong to the rank of devoutly religious men.' Quoted in: Frank, Phillip (1947) *Einstein: His Life and Times*, Chap. 12, section. 5.

13. Davies, P. (1992) *The Mind of God* provides a typical example of the awe expressed by cosmologists when observing the grandeur of the universe.

14. Lockley, M.G. (1999) *The Eternal Trail,* p. 296.

15. The famous 'God is Dead' quote comes from Friedrich Nietzsche, in *The Gay Science*, originally published in German in 1882.

16. The full quote from R. Dawkins' *The Blind Watchmaker,* is: 'Although atheism might have been *logically* tenable before Darwin, Darwin made it possible to be an intellectually fulfilled atheist.' (Dawkins R. 1986, p. 6.)

17. Rumi, J-D. (2001).

18. Hunt, H. (2006).

19. For citations on The 'Ambient Ecological Array of Perception' see Hunt (2006).

20. In many of his writings, Teilhard de Chardin frequently refers to the interiorness of matter

21. Jung, C.G. (1996) *The Psychology of Kundalini Yoga.*

22. Spinoza, B., Proposition 7, in *Ethics* Part II.

23. The Scientific and Medical Network holds regular (annual) Mystics and Scientists conferences.

24. See discussion of sociobiology in Chapter 2.

25. Thompson, W.I (1981) pp. 44-46.

26. Many scientists now admit convergence between creation and hero myth stories in archaic and modern (scientific) cultural settings.

27. Thompson, W.I. (1981) p. 63-64.

28. Forman, R.K.C. (1998).

29. Combs, A. (1996).

30. Krishna, G. (1967) *Kundalini: The Evolutionary Energy in Man*; Mookerjee, A. (1986) *Kundalini: The Arousal of the Inner Energ*; Sanella, L. (1987) *The Kundalini Experience*; White, J. (1990) *Kundalini: Evolution and Enlightenment.*

31. Oxtoby, Willard (1996/2001) *World Religions: Eastern Traditions.*

32. Saraswati, N.S. (2002) *Prana Pranayama, Prana Vidya.*

33. *Sakti* refers to the power of the Goddess said to lie like a coiled serpent at the base of the spine. When awakened it may rise to the crown chakra — to unite with the male supreme being known as the thousand petalled lotus.

34. Muller, F.M. (1873).

35. Bucke, R.M. (1901) *Cosmic Consciousness.*

36. Snow, P. (1991) *The Star Raft: Chinese Encounters with Africa.*

37. Havecker, C. (1987) *Understanding Aboriginal Culture.*

38. For discussion of Jacob Boehme, see Cheney, S. (1945) *Men Who Have Walked With God.* Other sources include *Interior Castle* by Teresa of Avila, and Merton, T. (1948) *The Seven Storey Mountain.*

39. Floor, cited in White, J. (1990) *Kundalini: Evolution and Enlightenment.*

40. The Hindu Vedas among the oldest known sacred texts dating to about 1500-1000 BC.

41. Saraswati, N.S. (2002) *Prana Pranayama, Prana Vidya.*

42. Rohen, J.W. (2007) *Functional Morphology.*

43. C.G. Jung (1996).

44. Fowler, B. (2000) *Iceman.*

45. Kaptchuk, T.J. (1983) *The Web That Has No Weaver: Understanding Chinese Medicine.*

46. Bentov, I. (1977) *Stalking The Wild Pendulum.* See also Chapter 3.

47. Ruse's comment appears on the cover of the McGrath team's reply: McGrath, A. and McGrath, J.C. (2007) *The Dawkins Delusion.* See Lockley M.G. (2007c). See also Ruse, M. (2008) and Chapter 7 above.

48. McGrath, A. (2005). Also McGrath, A. and McGrath, J.C. (2007). See note 47 above.

49. Lockley M.G. (2007).

50. McIntosh, S. (2007).

51. Polanyi, M. (1969) *Knowing and Being.*

52. Barfield, O. (1965) *Saving the Appearances: A Study in Idolatry.*

53. Goethe, J.W. von (1982) *Theory of Colours,* M.I.T. Press (originally published in German in 1810).

54. Sahtouris, E. (2000) *Earthdance.*

55. Loye, D. (2000) *Darwin's Lost Theory of Love.*

56. Armstrong, K. (2006) *The Great Transformation.* Also Thompson, W.I. (2004) *Self and Society: Studies in the Evolution of Consciousness.*

57. Clark, W.C. (1996) *Sex and the Origins of Death.*

58. Thesing, Carl (1933) *Genealogy of Love,* p. 1 and p. 245.

59. Malinowski, Bronislaw *The Sexual Life of Savages.*

60. Mellen, S.L. W. (1981) *The Evolution of Love.*

61. Hunt, M.M. (1959) *The Natural History of Love.*

62. Hunt, *op.cit..* p. 61.

63. Hunt, *op.cit..* p. 123.

64. Hunt, *op.cit..* p. 127.

65. Hunt, *op.cit..* p. 316.

66. Hunt, *op.cit..* p. 341.

67. Faulkner-Jones, 1982, p 28

68. Krishna, G. (1967) *Kundalini: The Evolutionary Energy in Man,* Shambhala, Boston. (See further in References and in note above.)

69. Bentov, I. (1977) p. 185.

70. Lorimer, D. (1990) *Whole in One.*

71. Teilhard De Chardin, *et al.* (1970) *Let Me Explain,* p. 64.

72. *Ibid.* p. 68.

73. Lemkow, A. (1990) *The Wholeness Principle.*

74. Teilhard de Chardin, P. (1965) *Building the Earth.*

75. Wilber, K. (1998) *The Marriage of Sense and Soul.*

76. While regarding Teilhard as a great mystic, and good scientist in many of his endeavours, Alister Hardy in *The Biology of God*, p. 208, rejects his thesis in *The Phenomenon of Man* as unscientific.

77. Tipler, F.J. (1994) *The Physics Of Immortality: Modern Cosmology, God and the Resurrection of the Dead.*

78. Teilhard de Chardin, P. (1975) *Toward the Future,* p. 86-87.

79. Rohen, J.W. (2007) *Functional Morphology.*

Postscript references

A 'slow movement towards morality', in Chesterton, G.K. (1919) *Orthodoxy,* pp. 198-202; '... Unless evils were allowed to break out ... , in Swedenborg, E. (1764) *Divine Providence* (#251); 'What academics call 'foundationalism...': for reference to 'foundationalism' in Descartes' *A Discourse on Method,* see McClaren, B.D. (2007); for 'The Buddha dug down...': see Thompson, W.I. (1996) p. 23.

References and Further Reading

Abbott, E. (1884) *Flatland: A Romance of Many Dimensions,* Seeley & Co, London.

Abbs, P. (1986) "The development of autobiography in western culture: from Augustine to Rousseau." Ph.D. thesis, University of Sussex, UK.

Abhayananda, S. (1996) *History of Mysticism: the Unchanging Testament,* third revised edition, Atma Books, Olympia WA.

Aczel, A.D. (2009) *The Cave and the Cathedral,* John Wiley and Sons.

Armande, J-C. (1974) *Gazelle Boy,* Bodley Head, London.

Armstrong, K. (2006) *The Great Transformation,* Atlantic Books.

Austin, J. (1998) *Zen and Brain,* MIT Press, Cambridge.

Badyaev, A.V. (2002) "Growing Apart: an ontogenetic perspective on the evolution of sexual size dimorphism." *Trends in Ecology and Evolution* 17: pp. 369–78.

Barfield, Owen (1926/1988) *History in English Words,* Lindisfarne Press.

—, (1965) *Saving the Appearances: Studies in Idolatry,* (second ed. 1988) Wesleyan University Press.

—, (1967) *Speaker's Meaning,* Wesleyan University Press.

Bell, E.T. (1951) *Mathematics: Queen and Servant of Science,* McGraw Hill.

Bell, J.S. (1964) "On the Einstein-Podolsky-Rosen Paradox", *Physics* 1, pp. 195–200.

Bentov, I. (1977) *Stalking the Wild Pendulum: On the Mechanics of Consciousness,* Destiny Books, Rochester, VT.

Bergson, H. (1988) *Matter and Memory,* Zone Books, New York.

Berman, M. (1989) *Coming To Our Senses: Body And Spirit in the Hidden History of the West,* Simon and Schuster, New York.

Bird, R.J. (2004) *Chaos and Life: Complexity and Randomness in Evolution and Thought,* Columbia University Press.

Blaxland de Lange, S. (2006) *Owen Barfield: Romanticism Come of Age — A Biography,* Temple Lodge Publications, London.

Blum, H.F. (1951) *Time's Arrow and Evolution,* Princeton University Press.

Bockemühl, J. (1985) ed. *Toward a Phenomenology of the Etheric World,* Anthroposophic Press, New York. Trans. from the German *Erscheinungsformen des Aetherischen* (1977), Freies Geistesleben, Stuttgart, Germany.

Bohm, D. (1980) *Wholeness and the Implicate Order,* Ark Paperbacks.

Bolk L. (1918), *Hersenen en Cultur,* Scheltema and Holkema, Amsterdam.

Bortoft, H. (1996) *The Wholeness of Nature: Goethe's Way of Science,* Lindisfarne Press, New York; Floris Books, Edinburgh, UK.

Brockman, J. (1995) *The Third Culture: Beyond the Scientific Revolution,* Simon and Schuster.

Bronowski, J. (1965) *William Blake and the Age of Revolution,* Harper and Row.

Brooke, J.H. (1991) *Science and Religion,* Cambridge University Press.

Bucke, R.M. (1901) *Cosmic Consciousness,* Dutton, New York.

Buss, D.M. (2004) *Evolutionary Psychology: The New Science Of The Mind,* Pearson, Boston.

Cahill, T. (1996) *How The Irish Saved Civilization*, Anchor Books.

Campbell, J. (1949) *The Hero With a Thousand Faces*, Bollingen, New York.

Cassirer, E. (1944) *An Essay on Man*, Yale University Press.

—, (1946) *Language and Myth*, Harper, New York.

Cheney, S. (1945) *Men who Have Walked with God*, Knopf, New York.

Chesterton, G.K. (1919) *Orthodoxy*, John Lane, London.

—, (1957) *St Francis of Assisi*, Image Books, UK.

Chopra, D. (2005) *Peace is the Way*, Harmony Books.

Clark, W.C. (1996) *Sex and the Origins of Death*, Oxford University Press.

Claxton, G. (1997) *Hare Brain, Tortoise Mind: Why Intelligence Increases When You Think Less*, HarperCollins.

Clottes, J. and Lewis Williams, D. (1998) *The Shamans of Prehistory: Trance and Magic in the Painted Caves*, H.N. Abrams, New York.

Coles, R. (1990) *The Spiritual Life of Children*, Houghton Mifflin, US.

Combs, A. (1996) *The Radiance of Being. Complexity, Chaos and the Evolution of Consciousness*, Paragon House, USA/ Floris Books, Edinburgh, UK. Revised ed. 2002, Paragon House.

Conan Doyle, A. (1921) *The Lost World*, Hodder and Stoughton, London.

Conway-Morris, S. (1998) *The Crucible of Creation*, Oxford University Press.

Crawford, C. and Krebs D. (2008) *Foundations of Evolutionary Psychology*, Routledge.

Curd, P. (1995) ed. *A Presocratics Reader*, Hackett Publishing Company, Cambridge, Mass.

Davies, P. (1992) *The Mind of God*, Simon and Schuster, New York.

Dawkins, R. (1976) *The Selfish Gene*, Oxford University Press.

—, (1986) *The Blind Watchmaker*, Norton, New York.

—, (2006) *The God Delusion*, Houghton Mifflin.

Demasio, R.A. (1994) *Descartes' Error: Emotion, Reason, and the Human Brain*, Quill, New York.

Dennett, D.C. (1991) *Consciousness Explained*, Back Bay Books, New York.

Descartes, René (1988) *The Philosophical Writings Of Descartes*, 3 vols., trans. John Cottingham, Robert Stoothoff, and Dugald Murdoch, Cambridge University Press, UK.

Dobbs, B.J.T (1975) *The Foundations of Newton's Alchemy*, Cambridge University Press.

Douglas-Smith, B. (1971) "An empirical study of religious mysticism", *British Journal of Psychiatry*, 118, pp. 549–54.

Du Nouy, L. (1947) *Human Destiny*, Longmans, Green and Co.

Eddington, A. (1964) *The Nature Of The Physical World*, Dent, London.

Edelglass, S. (2006) *The Physics of Human Experience*, Adonis Press,New York.

Edelman, G.M. and Tononi, G. (2000) *A Universe of Consciousness*, Basic Books, New York.

Einstein, A. (1961) *Relativity*, Bonanza Books, New York.

—, and Calaprice, A. (2000) eds. *The Expanded Quotable Einstein*, Princeton University Press.

Eisler, R. (1987) *The Chalice and the Blade*, Harper Collins.

Eliade, M. (1964) *Shamanism*, Princeton University Press.

Ellis, J. (1986) "The superstring: theory of everything, or of nothing?" *Nature* 323: pp. 595–98.

Endres, P. and Schad, W. (2002) *Moon Rhythms in Nature*, Floris Books, Edinburgh.

Erickson, E.H. (1959) "Identity and the Life Cycle: Selected Papers", *Psychological Issues*, Monograph 1, vol. 1, International Universities Press, New York.

—, (1968) *Identity: Youth and Crisis*, Norton, New York.

Fadiman, A. (1997) *The Spirit Catches You and You Fall Down,* The Noonday Press, New York.

Faulkner-Jones, D.E. (1982) *The English Spirit,* Rudolf Steiner Press, London.

Fawcett, D. (1916) *The World As Imagination,* Macmillan and Co. London.

Feliks, J. (2008) "Phi in the Acheulian", in Bednarik, R.G. and Hodgson, D. (eds.) *Pleistocene Paleoart of the World,* pp. 11–31, British Archaeological Reports International Series 1804, Oxford.

Feuerstein, G. (1987*)* *Structures of Consciousness: the Genius of Jean Gebser — an Introduction and Critique,* Integral Publishing, Lower Lake, CA.

Forman, R.K.C. (1998) "What does mysticism have to teach us about consciousness?" *Journal of Consciousness Studies 5*: pp. 185–201.

Fowler, B. (2000) *Iceman,* Random House.

Frank, Phillip (1947) *Einstein: His Life and Times.* Revised translation issued 1989/2002, Da Capo Press, US.

Freeman, A. (2003) *Consciousness: A Guide to the Debates,* ABC-Clio, Santa Barbara, CA.

Freud, Sigmund (1933) "New Introductory Lectures." In Vol. XXII, *The Complete Psychological Works of Sigmund Freud,* The Standard Edition, Hogarth Press, London (1964).

—, and Strachey, J. (ed.) (1960) *The Ego and the Id,* Standard Edition, W.W. Norton and Co., New York.

Frobenius, L. (1905) *Kulturgeschichte Afrikas,* Phaidon, Vienna, cited in Combs, A. (2002).

Fromm, E. (1941) *Fear of Freedom,* Routledge, UK, published in USA as *Escape from Freedom.*

Galbraith, J. (1998) "Spontaneous rising of Kundalini Energy: pseudo-schizophrenia or spiritual disease?" *Scientific and Medical Network Journal,* 67: p. 29.

Gebser, J. (1949/1986) *The Ever-Present Origin,* trans. Noel Barstad with Algis Mickunas, Ohio University Press, Athens, Ohio.

Ghose, A. and McDermott, R.A. (1994) *The Essential Aurobindo,* Steiner Books, New York.

Gilbert, S.R. (2000) *The Prophetic Imperative: Social Gospel in Theory and Practice,* Unitarian Universalist Association of Congregations, Boston, Mass.

Girard, R. (1986) *The Scapegoat,* John's Hopkins University Press, Baltimore.

Gleiser, M. (1997) *The Dancing Universe: From Creation Myths to the Big Bang,* Dartmouth College Press.

Goethe, J.W. (1795) *Erster Entwurf einer allgemeinen Einteitung in die vergleichende Anatomie, ausgehend von Osteologie,* in Steiner, R. (1883) *J.W. Goethe. Naturwissenschaftliche Schriften,* pp. 239–275. Also as Fotomechanischer Nachdruck (1982) R. Steiner Verlag, Dornach.

Goldstein, K. (1963) *The Organism: A Holistic Approach to Biology,* Beacon Press.

Goodenough, F.L. (1926) *Measurement of Intelligence by Drawing,* World Book Co., Chicago.

Goswami, A. (1995) *The Self-Aware Universe: How Consciousness Creates the Material World,* Tarcher/Putnam.

Gould, S.J. (1977a) *Ever Since Darwin: Reflections in Natural History,* Penguin.

—, (1977b) *Ontogeny and Phylogeny,* Harvard University Press, Cambridge.

—, (1985) "Geoffroy and the Homeobox," *Natural History,* 94: pp.12–23.

—, (1989) *Wonderful Life,* Norton, New York.

—, (2002) *The Structure of Evolutionary Theory,* Harvard University Press.

Grotstein, J. S. (1997) "Why Oedipus and not Christ?" *American Journal of Psychoanalysis,* 57: pp. 193–220 and 317–35.

Habermas, J. (1984–85) *The Theory Of Communicative Action,* (2 vols) trans. T.

McCarthy, Beacon, Boston.

Haeckel, E. (1866) *Generelle Morphologie der Organismen: allgemeine Grundzüge der organischen Formen-Wissenschaft, mechanisch begründet durch die von Charles Darwin reformirte Descendenz-Theorie,* 2 vols., George Reimer, Berlin.

Hapgood, C. (1964) *Maps of the Ancient Sea Kings: Evidence of Advanced Civilization in the Ice Age,* Dutton.

—, (1975) *Voices of Spirit,* Delacorte Press/Seymour Lawrence, New York.

Happold, F.C. (1963) *Mysticism: A Study and an Anthology,* Penguin.

Hardy, Alister (1975) *The Biology of God,* Jonathan Cape, London.

Harman, W. (1988) *Global Mind Change,* Knowledge Systems Inc.

Havecker, C. (1987) *Understanding Aboriginal Culture,* Cosmos Books, Sydney.

Hay, S.D. (2001) "The biological basis of spiritual awareness", in *Spirituality and Society in the new Millennium,* ed.Ursula King, Sussex Academic Press, pp. 124–35.

Ho, M-W. (1996) "The biology of free will", *Journal of Consciousness Studies,* 3: pp. 231–44.

—, (1998) *The Rainbow and the Worm: The Physics of Organisms,* World Scientific Publishing Co., Singapore.

Holbrook, B. (1981) *The Stone Monkey,* William Morrow, New York.

Holdredge, C. (1996) *Genetics and the Manipulation of Life,* Lindisfarne Press, Hudson, New York.

Hopson, J.A. (1980) "Relative brain size in dinosaurs — implications for dinosaurian endothermy", *American Association for the Advancement of Science Symposium* no. 28: pp. 287–310.

Hubbard, B.M. (1998) *Conscious Evolution,* New World Library.

Hunt, H. (2006) "The truth value of mystical experience", *Journal of Consciousness Studies* 13: pp. 5-43.

Hunt, M.M. (1959) *The Natural History of Love,* Knopf, US.

Huxley, A. (1945) *The Perennial Philosophy,* Harper and Brothers.

James, W. (1908) *The Varieties of Religious Experience,* Mentor Books.

Jaynes, Julian (1976) *The Origin of Consciousness in the Breakdown of the Bicameral Mind,* Houghton Mifflin, New York.

Jennings, H.S. (1930) The *Biological Basis of Human Nature,* Norton and Co., New York.

Jung, Carl Gustav (1931/1969) *The Structure of the Psyche,* in *The Collected Works of C. G. Jung,* Vol. 6, Princeton University Press.

—, (1961) *Memories, Dreams, Reflections,* Vintage Books.

—, (1966/1977) *Two Essays on Analytical Psychology,* in *The Collected Works of C. G. Jung,* Volume 7 (second edition) Princeton University Press, NJ.

—, (1972) *Synchronicity — An Acausal Connecting Principle,* Routledge and Kegan Paul.

—, (1976) *Psychological Types.* In *The Collected Works of C.G. Jung,* Vol.6, Princeton University Press, NJ.

—, (1981) *The Archetypes and the Collective Unconscious,* Princeton University Press, NJ.

—, (1996) *The Psychology of Kundalini Yoga,* Princeton University Press.

Kafatos, M. and Nadeau, R. (1990) *The Conscious Universe: Part and Whole in Modern Physical Theory,* Springer Verlag.

Kaptchuk, T.J. (1983) *The Web That Has No Weaver: Understanding Chinese Medicine,* Congdon and Weed Inc., New York.

Kawai, M. (1965) "Newly-acquired precultural behavior of the natural troop of Japanese monkeys on Koshima Island", in *Primates,* vol. 6, pp. 1–30.

Keck, L.R. (2000) The *Sacred Quest: The Evolution and Future of the Human Soul,* Chrysalis Books.

Keller, E.F. (1983) *A Feeling for the Organism,* W.H. Freeman and Company,

New York.

Kingsbury, B.F. (1924) "The significance of the so-called law of cephalocaudal differential growth", *Anatomical Record, v.* 27, pp. 305–21.

Koestler, A. (1959) *The Sleepwalkers,* Macmillan, New York.

—, (1969) "Beyond atomism and holism — the concept of the holon", pp. 192–232 in Koestler, A. and Smythies, J.R., eds., *Beyond Reductionism. New Perspectives in the Life Sciences,* New York, Macmillan.

Kohn, M. and Mithen, S.J. (1999) "Handaxes: products of sexual selection", *Antiquity,* 73: pp. 518–26.

König, K. (1969) *The First Three Years of the Child,* Anthroposophic Press, USA.

Krishna, G. (1967) *Kundalini: The Evolutionary Energy in Man,* Shambhala, Boston.

—, (1972) *The Secret of Yoga ,* Turnstone Press, Wellingborough.

Krishna, G. (1975) *The Awakening of the Kundalini,* Taraporevala Sons and Co., Mumbai, India.

—, (1997) *Kundalini: the evolutionary energy in man,* Shambhala, Boston, Mass.

Kühlewind, G. (1986) *The Logos-Structure of the World: Language as Model of Reality,* Lindisfarne Press, New York.

Kuhn, T. (1962) *The Structure of Scientific Revolutions,* University of Chicago Press.

Lakoff, G. and Johnson, M. (1999) *Philosophy in the Flesh,* Basic Books.

Laszlo, Ervin (1987) *Evolution: the Grand Synthesis,* Shambhala Books, Boston, Mass.

—, (2004) *Science and the Akashic Field: An Integral Theory of Everything,* Inner Traditions, Rochester, Vermont.

Lavie, P. (1996) *The Enchanted World of Sleep,* Yale University Press.

Le Guyader, H. (2004) *Geoffroy Saint-Hilaire: A Visionary Naturalist,* trans. M. Grene, University of Chicago Press, Chicago.

Lear, J. (1990/1998) *Love And Its Place In Nature: A Philosophical Interpretation of Freudian Psychoanalysis,* Yale University Press.

Lehrer, J. (2007) *Proust was a Neuroscientist,* Houghton Mifflin.

Lemkow, A.F. (1990) *The Wholeness Principle: Dynamics of Unity Within Science, Religion and Society,* Quest Books.

LeRoy Ladurie, E. (1980) *Montaillou: Cathars and Catholics in a French Village,* 1294–1324, Penguin.

Lockley, M.G. (1999) *The Eternal Trail: a Tracker Looks at Evolution,* Perseus Books, Reading, MA.

—, (2000) "A broader look at spiritual emergence experience: implications for consciousness studies", *Scientific and Medical Network Journal,* 72: p. 18–19.

—, (2005) "The intelligent universe paradigm: Red herrings and red heresies in the 'Intelligent Design' debate". *Scientific and Medical Network Journal,* 88: pp. 5–9.

—, (2007a) "How Western Science Views, Tackles and Constrains Consciousness Studies", *Scientific and Medical Network Journal,* 93: p. 57–58.

—, (2007b): "The morphodynamics of dinosaurs, other archosaurs and their trackways: holistic insights into relationships between feet, limbs and the whole body". In *Ichnology at the crossroads: a multidimensional approach to the science of organism-substrate interactions,* R. Bromley and R. Melchor (eds), Society of Economic and Paleontologists and Mineralogists Special Publication 88: pp. 27–51.

—, (2007c) "Dawkins Embarrasses Atheists", *Scientific and Medical Network Journal,* 95: p. 51–52.

Lodge, R.C. (1956) *The Philosophy of Plato,* Routledge, London.

Long, B. (1984) *The Origins of Man and the Universe: The Myth that Came to Life,* Routledge and Kegan Paul.

Lorimer, D. (1990) *Whole in One,* Arkana, Penguin.

Lovelock, J. (1979) *Gaia: A New Look At Life On Earth,* Oxford University Press.

Loye, D. (2000) *Darwin's Lost Theory of Love,* Excel Press, Lincoln, Nebraska.

MacLean, Paul D. (1990) *The Triune Brain In Evolution: Role In Paleocerebral Functions,* Plenum Press, New York.

Manchester, W. (1992) *A World Lit Only By Fire,* Little Brown and Co., New York.

Mandelbrot, B.B (1982) *The Fractal Geometry of Nature,* W.H. Freeman and Co., New York.

Margulis, L. (1970) *The Origin of Eukaryotic Cells,* Yale University Press, New Haven.

Marshack, A. (1972) *The Roots of Civilization,* McGraw Hill.

McClaren, B.D. (2007) *Everything Must Change,* T. Nelson.

McGrath, A. (2005) *Dawkins' God: Genes, Memes and the Meaning of Life,* Blackwell Publishing, Oxford.

—, and McGrath, J.C. (2007) *The Dawkins Delusion,* Society for Promoting Christian Knowledge.

McIntosh, S. (2007) *Integral consciousness and the Future of Evolution,* Paragon House.

McNamara, K. (1997) *Shapes of Time: the Evolution of Growth and Development,* Johns Hopkins University Press, Baltimore.

Medawar, P.B. (1996) *The Strange Case of the Spotted Mice, and Other Classic Essays on Science,* Oxford University Press.

Meister Eckhart (1909) *Sermons,* trans. C. Field. H.R. Allenson, London.

Mellen, S.L. W. (1981) *The Evolution of Love,* W.H Freeman and Co.

Merton, T. (1948) *The Seven Storey Mountain,* Harcourt Brace, New York.

Milner, R. (1990) *The Encyclopedia of Evolution,* Facts on File, New York.

Mishlove, J. (1975) *The Roots of Consciousness,* Random House.

Mithen, S.J. (1996) *The Prehistory of the Mind,* Thames and Hudson, London.

—, (2003) "Handaxes: the first aesthetic artifacts", in Eckhart, V. and Grammer,

K. (eds.) *Evolutionary Aesthetics,* pp. 261–75, Springer Verlag, Berlin.

Mookerjee, A. (1986) *Kundalini: The Arousal of the Inner Energy,* Destiny Books, Rochester.

Muller, F.M. (1873) *Lectures on the Origin and Growth of Religion as Illustrated by the Religions of India,* Charles Scribner's, New York.

Myers, D.G. (2007) *Psychology* (eighth edition, module 11) "Adulthood, and reflections on Developmental Issues," pp. 171–76,Worth Publishers, New York.

Needham, J. (1954–2008) *Science and Civilization in China,* 22 vols. Cambridge University Press.

Newberg, A., D'Aquili, E. and Rause, V. (2001) *Why God Won't Go Away: Brain Science and the Biology of Belief,* Ballantine Books.

Nitobé, I. (1998) *Bushido, the Soul of Japan,* Tokuhei Suchi, Tokyo.

Otto, R. (1969) *The Idea of the Holy,* Oxford University Press.

Oxtoby, Willard (1996/2001) *World Religions: Eastern Traditions,* Oxford University Press.

Pagels, E. (1979) *The Gnostic Gospels,* Vintage Books.

Pals, D.L. (1996) *Seven Theories of Religion,* Oxford University Press.

Penrose, R. (1989) *The Emperor's New Mind.* Oxford University Press.

Piaget, J. (1976) *The Grasp Of Consciousness: Action and Concept In the Young Child,* Harvard University Press.

Plato (1966) *Phaedo,* trans. Harold North Fowler, in *Plato in Twelve Volumes,* Vol.1, Loeb Classical Series, Harvard University Press, Cambridge, MA; William Heinemann, London.

Plotinus (1991) *The Enneads,* trans. S. MacKenna and J. Dillon, Penguin Books, New York.

Polanyi, M. (1964) *Science, Faith and Society,* University of Chicago Press.

—, (1969) *Knowing and Being,* (ed. M. Grene) University of Chicago Press.

Pollack, J.H. (1965) *Croiset the Clairvoyant,* Bantam Books, New York.

Portmann, A. (1964) *New Paths in Biology,* Harper Row , New York.

Priestly, J.B. (1937) *Midnight on the Desert,* Harper.

Radin, D.I (1997) *The Conscious Universe,* Harper Edge.

Ravindra, R. (1998) *Christ the Yogi: A Hindu Reflection on the Gospel of John.* Revised edition 2004, published as *The Gospel of John in the Light of Indian Mysticism,* Inner Traditions, Rochester, Vermont.

Riegner, M. (1985) "Horns, spots and stripes: form and pattern in mammals." *Orion Nature Quarterly,* 4(4): pp. 22–35.

—, (2008) "Parallel evolution of plumage pattern and coloration in birds: implications for defining avian morphospace." *The Condor,* 110: pp. 599–614.

Rightmire, G.P. (2004) "Brain size and encephalization in early to Mid-Pleistocene Homo", *American Journal of Physical Anthropology,* 124 (2), p. 109.

Rohen, J.W. (2007) *Functional Morphology. The Dynamic Wholeness of the Human Organism,* Adonis Press.

Rumi, J-D. (2001) *The Selected Poems of Rumi,* trans. R.A. Nicholson, Dover Publications.

Ruse, M. (2008) *Evolution and Religion: A Dialog,* Rowman and Littlefield, New York.

Russell, Peter (1995) *The Global Brain,* Global Brain Inc, Palo Alto. Reissued as *The Global Brain: The Awakening Earth in a New Century,* Floris Books, Edinburgh, 2007.

Ruthven, M. (1997) *Islam: A Very Short Introduction,* Oxford University Press.

Sahtouris, E. (2000) *Earthdance,* iUniverse Inc., New York.

Saint-Hilaire, G. (1822) "Considérations générales sur la vertèbre", *Memoir du Musée National de l'Histoire Naturelle* 9.

Sannella, L. (1987) *The Kundalini Experience,* Integral Publishing, Lower Lake, CA.

Santillana, G. de and von Dechend, H. (1969) *Hamlet's Mill,* Gambit, Boston.

Saraswati, N.S. (2002) *Prana Pranayama, Prana Vidya,* Yoga Publications Trust, Mungar, Bihar, India.

Schad, W. (1977) *Man and Mammals: Towards a Biology of Form,* Waldorf Press, New York.

Schopf, T.J.M. (1972) ed., *Models in Paleobiology,* Freeman Cooper, San Francisco.

Searle, J. (1984/2003) *Minds, Brains, and Science,* Harvard University Press, Cambridge.

—, (2004) *Mind: A Brief Introduction,* Oxford University Press.

Shelden, W.H. (1944) *The Varieties of Temperament,* Harper Brothers.

Sheldrake, R. (1981) *A New Science Of Life,* HarperCollins.

—, (1994) *The Rebirth of Nature: The Greening of Science and God,* Inner Traditions, Rochester, VT., p. 260.

—, (1999) *Dogs That Know When Their Owners Are Coming Home,* Crown Publishers, New York.

Shepherd, A.P. (1983) *Scientist Of The Invisible: Rudolf Steiner,* Inner Traditions, Rochester VT; Floris Books, Edinburgh, UK.

Sherrard P. (1992) *Human Image, World Image,* Golgonooza Press, Ipswich, UK.

Slotten, R.A. (2004) *The Heretic In Darwin's Court: The Life of Alfred Russel Wallace,* Columbia University Press.

Snow, C.P. (1959) *The Two Cultures and Scientific Revolution,* Cambridge University Press.

Snow, P. (1991) *The Star Raft: Chinese Encounters with Africa,* Weidenfeld and Nicholson, UK.

Spetner, L.M. (1997) *Not by Chance,* Judaica Press, Brooklyn, NY.

Spinoza, B. de (1982) *The Ethics and Selected Letters,* trans. Samuel Shirley, Hackett Publishing, Indianapolis.

Stace, W.T. (1960) *The Teachings of the*

Mystics, Mentor Books, New York.

Steinbock, A.J. (2007) *Phenomenology and Mysticism,* Indiana University Press.

Steiner, Heiri and Gebser, Jean (1962) *Anxiety: A Condition of Modern Man,* trans. Peter Heller, Dell Publishing, New York.

Steiner, Rudolf (1961) *Christianity as Mystical Fact,* Rudolf Steiner Publications, New York.

—, (1963) *The Life, Nature and Cultivation of Anthroposophy,* Rudolf Steiner Press, London

—, (1971) *Cosmic Memory,* Rudolf Steiner Publications, New York.

—, (1999) *A Psychology of Body, Soul and Spirit.* Trans. Marjorie Spock. Steiner Books, New York.

Strauss, W. and Howe, N. (1997) *The Fourth Turning,* Broadway Books.

Sugerman, S. (1976) ed. *Evolution of Consciousness,* Wesleyan University Press.

Swedenborg, E. (1764) *The Wisdom of Angels. Concerning the Divine Providence,* Amsterdam.

Teilhard de Chardin, P. (1959) *The Phenomenon of Man,* Harper Row, London.

—, (1964) *The Future of Man,* Harper, New York.

—, (1965) *Building the Earth,* Dimension Books, Wilkes-Barre, Pennsylvania.

—, (1970) with Demoulin, J-P. and Hague, R. *Let Me Explain,* Collins Fontana, UK.

—, (1975) *Toward the Future,* Collins, London.

Thesing, C. (1933) *Genealogy of Love,* Routledge and Sons, London.

Thompson, W.I. (1973) *Passages about Earth,* Harper and Row.

—, (1981) *Time Falling Bodies Take to Light,* St Martin's Press, New York.

—, (1996) *Coming into Being,* St Martin's Press, New York.

—, (2004) *Self and Society: Studies in the Evolution of Consciousness,* Imprint Academic, Exeter, UK.

Tipler, F.J. (1994) *The Physics Of Immortality: Modern Cosmology, God and the Resurrection of the Dead,* Doubleday.

Toumlin, S. and Goodfield, J. (1967) *The Discovery Of Time,* Penguin.

Underhill, E. (1911) *Mysticism,* Methuen, London. Reissued Bracken Books, 1995.

Usener, Hermann (1896) *Götternamen: Versuch einer Lehre von der religiösen Begriffsbildung.* Third edition 1948. Reissued 2000 by Vittorio Klostermann, Germany.

Van der Post, L. (1958) *The Lost World of the Kalahari,* Hogarth Press, UK.

Verhulst, J. (2003) *Developmental Dynamics in Humans and other Primates: Discovering Evolutionary Principles Through Comparative Morphology,* Adonis Press, New York.

Vernadsky, Vladimir (1945) "The Biosphere and Noosphere", *American Scientist* 33: pp. 1–12.

Verne, J. (1870) *Vingt Mille Lieues Sous Les Mers,* P. J Hetzel, Paris. First published in English in 1872, as *20,000 leagues Under the Sea.*

Vico, G. (1725/1994) *The Life of Giambattista Vico written by himself,* trans. M.H. Fisch and T.H. Bergin, Cornell University Press.

Vitaliano, G. (2000) "A new integrative model for states of consciousness." *NLP World* 7: pp. 41–82.

Weiss, B.L. (1988) *Many Lives, Many Masters,* Simon and Schuster,New York.

Weiss, P. (1973) *The Science of Life,* Futura Publishing, New York.

Whincup, P.H., and J.A. Gilg, K. Odoki, S.J.C. Taylor, D.G. Cook (2001) "Age of menarche in contemporary British teen-agers: survey of girls born between 1982 and 1986", *British Medical Journal,* vol. 22, p. 1095–96.

White, J. (1990) *Kundalini: Evolution and Enlightenment,* Paragon House, New York.

Wilber, K. (1977) *The Spectrum of Consciousness,* A Quest book,

Theosophical Publishing House, Wheaton, Illinois.

—, (1979) *No Boundary*, Shambhala, Boston.

—, (1995) *Sex, Ecology and Spirituality: The Spirit of Evolution*, Shambhala, Boston.

—, (1997) "An integral theory of consciousness", *Journal of Consciousness Studies* 4, pp. 71–92.

—, (1998) *The Marriage of Sense and Soul*, Random House.

Wilkinson, R. (1996) *The Spiritual Basis of Steiner Education*, Sophia Books.

Wilson, E.O. (1975) *Sociobiology, The New Synthesis*, Harvard University Press.

—, (1999) *Consilience*, Vintage Books.

Yoke, H.P. (1985) *Li, Qi and Shu: An Introduction to Science and Civilization in China*, Univ. of Washington Press, Seattle.

Young, D. (1991) *Origins of the Sacred: The Ecstasies of Love and War*, St Martin's Press, New York.

Yutang, L. (1935) *My Country and My People*, Reynal and Hitchcock, New York.

Zhang, C. (2003) "Electromagnetic Body v. Chemical Body", *Scientific and Medical Network Journal*, 81: pp. 7–10.

Index